Aspects of Differential Geometry II

Synthesis Lectures on Mathematics and Statistics

Editor
Steven G. Krantz, ***Washington University, St. Louis***

Aspects of Differential Geometry II
Peter Gilkey, JeongHyeong Park, and Ramón Vázquez-Lorenzo
2015

Aspects of Differential Geometry I
Peter Gilkey, JeongHyeong Park, and Ramón Vázquez-Lorenzo
2015

An Easy Path to Convex Analysis and Applications
Boris S. Mordukhovich and Nguyen Mau Nam
2013

Applications of Affine and Weyl Geometry
Eduardo García-Río, Peter Gilkey, Stana Nikčević, and Ramón Vázquez-Lorenzo
2013

Essentials of Applied Mathematics for Engineers and Scientists, Second Edition
Robert G. Watts
2012

Chaotic Maps: Dynamics, Fractals, and Rapid Fluctuations
Goong Chen and Yu Huang
2011

Matrices in Engineering Problems
Marvin J. Tobias
2011

The Integral: A Crux for Analysis
Steven G. Krantz
2011

Statistics is Easy! Second Edition
Dennis Shasha and Manda Wilson
2010

Lectures on Financial Mathematics: Discrete Asset Pricing
Greg Anderson and Alec N. Kercheval
2010

Jordan Canonical Form: Theory and Practice
Steven H. Weintraub
2009

The Geometry of Walker Manifolds
Miguel Brozos-Vázquez, Eduardo García-Río, Peter Gilkey, Stana Nikčević, and Ramón Vázquez-Lorenzo
2009

An Introduction to Multivariable Mathematics
Leon Simon
2008

Jordan Canonical Form: Application to Differential Equations
Steven H. Weintraub
2008

Statistics is Easy!
Dennis Shasha and Manda Wilson
2008

A Gyrovector Space Approach to Hyperbolic Geometry
Abraham Albert Ungar
2008

Aspects of Differential Geometry II
Peter Gilkey, JeongHyeong Park, and Ramón Vázquez-Lorenzo

ISBN: 978-3-031-01280-8 paperback
ISBN: 978-3-031-02408-5 ebook

DOI 10.1007/978-3-031-02408-5

A Publication in the Springer series
SYNTHESIS LECTURES ON MATHEMATICS AND STATISTICS

Lecture #16
Series Editor: Steven G. Krantz, *Washington University, St. Louis*
Series ISSN
Print 1938-1743 Electronic 1938-1751

Aspects of Differential Geometry II
Peter Gilkey, JeongHyeong Park, and Ramón Vázquez-Lorenzo

ISBN: 978-3-031-01280-8 paperback
ISBN: 978-3-031-02408-5 ebook

DOI 10.1007/978-3-031-02408-5

A Publication in the Springer series
SYNTHESIS LECTURES ON MATHEMATICS AND STATISTICS

Lecture #16
Series Editor: Steven G. Krantz, *Washington University, St. Louis*
Series ISSN
Print 1938-1743 Electronic 1938-1751

Statistics is Easy! Second Edition
Dennis Shasha and Manda Wilson
2010

Lectures on Financial Mathematics: Discrete Asset Pricing
Greg Anderson and Alec N. Kercheval
2010

Jordan Canonical Form: Theory and Practice
Steven H. Weintraub
2009

The Geometry of Walker Manifolds
Miguel Brozos-Vázquez, Eduardo García-Río, Peter Gilkey, Stana Nikčević, and Ramón Vázquez-Lorenzo
2009

An Introduction to Multivariable Mathematics
Leon Simon
2008

Jordan Canonical Form: Application to Differential Equations
Steven H. Weintraub
2008

Statistics is Easy!
Dennis Shasha and Manda Wilson
2008

A Gyrovector Space Approach to Hyperbolic Geometry
Abraham Albert Ungar
2008

Aspects of Differential Geometry II

Peter Gilkey
University of Oregon, Eugene, OR

JeongHyeong Park
Sungkyunkwan University, Suwon, Korea
Institute for Advanced Study, Seoul, Korea

Ramón Vázquez-Lorenzo
University of Santiago de Compostela, Santiago de Compostela, Spain

SYNTHESIS LECTURES ON MATHEMATICS AND STATISTICS #16

ABSTRACT

Differential Geometry is a wide field. We have chosen to concentrate upon certain aspects that are appropriate for an introduction to the subject; we have not attempted an encyclopedic treatment.

Book II deals with more advanced material than Book I and is aimed at the graduate level. Chapter 4 deals with additional topics in Riemannian geometry. Properties of real analytic curves given by a single ODE and of surfaces given by a pair of ODEs are studied, and the volume of geodesic balls is treated. An introduction to both holomorphic and Kähler geometry is given. In Chapter 5, the basic properties of de Rham cohomology are discussed, the Hodge Decomposition Theorem, Poincaré duality, and the Künneth formula are proved, and a brief introduction to the theory of characteristic classes is given. In Chapter 6, Lie groups and Lie algebras are dealt with. The exponential map, the classical groups, and geodesics in the context of a bi-invariant metric are discussed. The de Rham cohomology of compact Lie groups and the Peter–Weyl Theorem are treated. In Chapter 7, material concerning homogeneous spaces and symmetric spaces is presented. Book II concludes in Chapter 8 where the relationship between simplicial cohomology, singular cohomology, sheaf cohomology, and de Rham cohomology is established.

We have given some different proofs than those that are classically given and there is some new material in these volumes. For example, the treatment of the total curvature and length of curves given by a single ODE is new as is the discussion of the total Gaussian curvature of a surface defined by a pair of ODEs.

KEYWORDS

Chern classes, Clifford algebra, connection, de Rham cohomology, geodesic, Jacobi operator, Kähler geometry, Levi–Civita connection, Lie algebra, Lie group, Peter–Weyl Theorem, pseudo-Riemannian geometry, Riemannian geometry, sheaf cohomology, simplicial cohomology, singular cohomology, symmetric space, volume of geodesic balls

This book is dedicated to
Alison, Arnie, Carmen, Junmin,
Junpyo, Manuel, Montse, Rosalía, and Susana.

Contents

Preface

This two-volume series arose out of work by the three authors over a number of years both in teaching various courses and also in their research endeavors.

The present volume (Book II) is comprised of five chapters that continue the discussion of Book I. In Chapter 4, we examine the geometry of curves which are the solution space of a constant coefficient ordinary differential equation. We give necessary and sufficient conditions that the curves give a proper embedding and we examine when the total extrinsic curvature is finite. We examine similar questions for the total Gaussian curvature of a surface defined by a pair of ODEs and apply the Gauss–Bonnet Theorem to express the total Gaussian curvature in terms of the curves associated to the individual ODEs. We then examine the volume of a small geodesic ball in a Riemannian manifold. We show that if the scalar curvature is positive, then volume grows more slowly than it does in flat space while if the scalar curvature is negative, then volume grows more rapidly than it does in flat space. Chapter 4 concludes with a brief introduction to holomorphic and Kähler geometry.

Chapter 5 treats de Rham cohomology. The basic properties are introduced and it is shown that de Rham cohomology satisfies the Eilenberg–Steenrod axioms; these are properties that all homology and cohomology theories have in common. We shall postpone until Chapter 8 a discussion of the Mayer–Vietoris sequence and the homotopy property as these depend upon some results in homological algebra that will be treated there. We determine the de Rham cohomology of the sphere and of real projective space. We introduce Clifford algebras and present the Hodge Decomposition Theorem. This is used to establish the Künneth formula and Poincaré duality. We treat the first Chern class in some detail and use it to determine the ring structure of the de Rham cohomology of complex projective space. A brief introduction to the higher Chern classes and the Pontrjagin classes is given.

Chapter 6 contains an introduction to the theory of Lie groups and Lie algebras. We restrict for the most part to matrix groups so that the exponential and log functions can be given explicitly in terms of convergent power series. We show in this setting that a closed subgroup of a matrix group is a Lie subgroup; this result is used to treat the geometry of the classic matrix groups (special linear group, orthogonal group in arbitrary signature, unitary group in arbitrary signature, symplectic group, etc.). If M is a compact Lie group with a bi-invariant metric, we show the one-parameter subgroups of the exponential map are geodesics. We express the de Rham cohomology of a compact connected Lie group in terms of the left-invariant differential forms and prove the Hopf structure theorem that shows the de Rham cohomology of a compact connected Lie group is a finitely generated exterior algebra on odd-dimensional generators. We use these results to determine the ring structure of the de Rham cohomology of the unitary group.

We conclude Chapter 6 by discussing the orthogonality relations and the Peter–Weyl Theorem which decomposes L^2 as the direct sum of irreducible representations for a compact connected Lie group.

Chapter 7 presents an introduction to the theory of homogeneous spaces and of symmetric spaces. We examine coset spaces and the group of isometries of a pseudo-Riemannian manifold. We introduce material related to the Lie derivative and Killing vector fields. We outline the geometry of homogeneous spaces, of local symmetric spaces, and of global symmetric spaces. Chapter 8 concludes Book II with a discussion of other cohomology theories. We introduce the necessary homological machinery to show that de Rham cohomology is a homotopy functor and has the Mayer–Vietoris long exact sequence; these two results also have a significant geometric input. We relate simplicial cohomology, singular cohomology, and sheaf cohomology to de Rham cohomology.

We have tried whenever possible to give the original references to major theorems in this area. We have provided a number of pictures to illustrate the discussion. Chapters 1 and 2 of Book I are suitable for an undergraduate course on "Calculus on Manifolds" and arose in that context out of a course at the University of Oregon. Chapter 3 is designed for an undergraduate course in Differential Geometry. Therefore, Book I is suitable as an undergraduate text although, of course, it also forms the foundation of a graduate course in Differential Geometry as well. Book II can be used as a graduate text in Differential Geometry and arose in that context out of second year graduate courses in Differential Geometry at the University of Oregon and at Sungkyunkwan University. The material can, however, also form the basis of a second semester course at the undergraduate level as well. While much of the material is, of course, standard, many of the proofs are a bit different from those given classically and we hope provide a new viewpoint on the subject. Our treatment of curves in $\mathbb{R}^m$ given by the solution to constant coefficient ODEs which have finite total curvature is new as is the corresponding treatment of the total Gaussian curvature of a surface given by a pair of ODEs. There are other examples; Differential Geometry is of necessity a vibrant and growing field; it is not static! There are, of course, many topics that we have not covered. This is a work on "Aspects of Differential Geometry" and of necessity must omit more topics than can be possibly included.

For technical reasons, the material is divided into two books and each book is largely self-sufficient. To facilitate cross references between the two books, we have numbered the chapters of Book I from 1 to 3, and the chapters of Book II from 4 to 8.

Peter Gilkey, JeongHyeong Park, and Ramón Vázquez-Lorenzo
May 2015

Acknowledgments

We have provided many images of famous mathematicians in these two books; mathematics is created by real people and we think having such images makes this point more explicit. The older pictures are in the public domain. We are grateful to the Archives of the Mathematisches Forschungsinstitut Oberwolfach for permitting us to use many images from their archives (R. Brauer, H. Cartan, S. Chern, G. de Rham, S. Eilenberg, H. Hopf, E. Kähler, H. Künneth, L. Nirenberg, H. Poincaré, W. Rinow, L. Vietoris, and H. Weyl); the use of these images was granted to us for the publication of these books only and their further reproduction is prohibited without their express permission. Some of the images (E. Beltrami, E. Cartan, G. Frobenius, and F. Klein) provided to us by the MFO are from the collection of the Mathematische Gesellschaft Hamburg; again, the use of any of these images was granted to us for the publication of these books only and their further reproduction is prohibited without their express permission.

The research of the authors was supported by the Basic Science Research Program through the National Research Foundation of Korea (NRF) funded by the Ministry of Education (2014053413) and by Project MTM2013-41335-P with FEDER funds (Spain). The authors are very grateful to Esteban Calviño-Louzao and Eduardo García-Río for constructive suggestions and assistance in proofreading. The assistance of Ekaterina Puffini of the Krill Institute of Technology has been invaluable. Wikipedia has been a useful guide to tracking down the original references and was a source of many of the older images that we have used that are in the public domain.

Peter Gilkey, JeongHyeong Park, and Ramón Vázquez-Lorenzo
May 2015

CHAPTER 4

Additional Topics in Riemannian Geometry

We continue the discussion of various topics in Riemannian geometry which was started in Chapter 3 of Book I. In Section 4.1, we discuss the geometry of curves and surfaces in Euclidean space that arise as solutions to an ordinary differential equation (ODE); this involves a nice application of the Gauss–Bonnet Theorem that is typical in the subject. In Section 4.2, we relate the scalar curvature to the growth of the volume of geodesic balls. In Section 4.3, we turn to holomorphic geometry. We present the Cauchy–Riemann equations, define almost complex structures, and state the Newlander–Nirenberg Theorem. We decompose $d = \partial + \bar{\partial}$ and introduce the spaces $\Lambda^{p,q}$. We define complex projective space. In Section 4.4, we present an introduction to Kähler geometry.

4.1 CURVES AND SURFACES IN $\mathbb{R}^n$ GIVEN BY ODES

Let P be a constant coefficient ODE. The solution set of P defines a real analytic curve $\sigma_P(t)$. Given two such ODEs, let $\Sigma_{P_1,P_2}(t_1,t_2) := \sigma_{P_1}(t_1) \otimes \sigma_{P_2}(t_2)$ be a real analytic surface. In Section 4.1, we shall present some results of Gilkey et al. [21, 22] dealing with the total first curvature $\kappa[\sigma_P]$ of the curve σ_P and the total Gaussian curvature $K[\Sigma_{P_1,P_2}]$ of the surface Σ_{P_1,P_2}. The study of $K[\Sigma_{P_1,P_2}]$ illustrates the manner in which the Gauss–Bonnet Theorem is often used.

Let $P(\phi) := \phi^{(n)} + c_{n-1}\phi^{(n-1)} + \cdots + c_0\phi$ for $c_i \in \mathbb{R}$ be a constant coefficient ODE. Let $\mathcal{P} = \mathcal{P}_P$ be the associated characteristic polynomial and let $\mathcal{R} = \mathcal{R}_P$ be the roots of $\mathcal{P}$:

$$\mathcal{P}(\lambda) := \lambda^n + c_{n-1}\lambda^{n-1} + \cdots + c_0 \quad \text{and} \quad \mathcal{R} := \{\lambda \in \mathbb{C} : \mathcal{P}(\lambda) = 0\}.$$

We suppose that all the roots of $\mathcal{P}$ have multiplicity 1; the extension to the higher multiplicity case is for the most part straightforward modulo adding suitable powers of t. Enumerate the roots $\mathcal{R}$ of $\mathcal{P}$ in the form:

$$\mathcal{R} = \{r_1, \ldots, r_k, z_1, \bar{z}_1, \ldots, z_u, \bar{z}_u\} \quad \text{for} \quad k + 2u = n,$$

where $r_i \in \mathbb{R}$ for $1 \le i \le k$ and where $z_j = a_j + \sqrt{-1}b_j$ with $b_j > 0$ for $1 \le j \le u$. Choose the labeling so $r_1 > r_2 > \cdots$ and $a_1 \ge a_2 \ge \cdots$. The standard basis for the solution space of P

is given by the real analytic functions

$$\begin{array}{lll} \phi_1 := e^{r_1 t}, & & \dots, \quad \phi_k := e^{r_k t}, \\ \phi_{k+1} := e^{a_1 t}\cos(b_1 t), & \phi_{k+2} := e^{a_1 t}\sin(b_1 t), & \dots, \\ \phi_{n-1} := e^{a_u t}\cos(b_u t), & \phi_n := e^{a_u t}\sin(b_u t). & \end{array}$$

The associated curve $\sigma_P : \mathbb{R} \to \mathbb{R}^n$, the curvature κ of σ_P, the element of arc length ds, and the total first curvature $\kappa[\sigma_P]$ are given by:

$$\sigma_P(t) := (\phi_1(t), \dots, \phi_n(t)), \quad \kappa := \frac{\|\dot\sigma_P \wedge \ddot\sigma_P\|}{\|\dot\sigma_P\|^3},$$
$$ds := \|\dot\sigma_P\| dt, \quad \kappa[\sigma_P] := \int_{\sigma_P} \kappa ds = \int_{-\infty}^{\infty} \frac{\|\dot\sigma_P \wedge \ddot\sigma_P\|}{\|\dot\sigma_P\|^2} dt.$$

Let $\Re(\lambda)$ (resp. $\Im(\lambda)$) be the real (resp. imaginary) part of a complex number λ. We say the *real roots are dominant* if $r_1 > 0 > r_k$ and if $r_1 > \Re(\lambda) > r_k$ for any $\lambda \in \mathcal{R} - \{r_1, r_k\}$. This means that $\lim_{t\to\infty} \|e^{-r_1 t}\sigma_P\| = 1$, $\lim_{t\to-\infty} \|e^{-r_k t}\sigma_P\| = 1$, and $\lim_{t\to\pm\infty} \|\sigma_P\| = \infty$. The remaining roots are then said to be *subdominant*. We refer to P. Gilkey, C. Y. Kim, H. Matsuda, J. H. Park, and S. Yorozu [21] for the proof of the following result.

C. Y. Kim

H. Matsuda

S. Yorozu

Theorem 4.1 Let σ_P be the curve defined by a real constant coefficient ordinary differential equation P of order n with simple roots and with $r_1 > 0 > r_k$.

1. The curve σ_P is a proper embedding of $\mathbb{R}$ into $\mathbb{R}^n$ with infinite length. The total first curvature $\kappa[\sigma_P]$ is finite if and only if the real roots are dominant.
2. If there are no complex roots, then $\kappa[\sigma_P] \le \frac{\pi}{4} n(n-1)$. Given $\epsilon > 0$, there exists an ODE P_ϵ of order n with no complex roots so that $\kappa[\sigma_{P_\epsilon}] \ge \frac{1}{3}(n-1) - \epsilon$ and so that the real roots are dominant. Consequently, any universal upper bound must grow at least linearly with n and at worst quadratically. If subdominant complex roots are allowed, then there is no uniform upper bound.

If σ_P is defined by some other basis for the solution space of P than the standard basis, then Assertion 1 continues to hold and can be generalized to the case when the roots have multiplicities greater than 1.

Given a constant coefficient ODE P_1 (resp. P_2) of order n_1 (resp. n_2), we can define $\Sigma(t_1, t_2) := \sigma_{P_1}(t_1) \otimes \sigma_{P_2}(t_2)$ mapping $\mathbb{R}^2$ to $\mathbb{R}^{n_1 n_2}$. If $\{\phi_i\}$ is the standard basis for the solution space of P_1 and if $\{\psi_j\}$ is the standard basis for the solution space of P_2, then $\{\phi_i(t_1)\psi_j(t_2)\}$ are the coordinates of $\Sigma(t_1, t_2)$. Theorem 4.1 extends to this setting where we replace κ by the Gaussian curvature K. The total absolute Gaussian curvature is defined to be $\int_\Sigma |K|\,|\,\mathrm{dvol}\,|$. We refer to Gilkey et al. [22] for the proof of the following result.

Theorem 4.2 *Adopt the notation established above.*

1. *If the real roots of $\mathcal{P}_1$ and $\mathcal{P}_2$ are dominant, then Σ is a geodesically complete proper embedding of $\mathbb{R}^2$ into $\mathbb{R}^{n_1 n_2}$ of finite total absolute Gaussian curvature which has infinite volume.*

2. *Let H be the mean curvature vector. Assume all the roots of $\mathcal{P}_1$ (resp. $\mathcal{P}_2$) are real and that there are at least two positive and at least two negative roots. Then there exist $\epsilon = \epsilon(\Sigma) > 0$ and $C = C(\Sigma) > 0$ so that $\|H\| \le Ce^{-\epsilon\|(t_1,t_2)\|}$. Furthermore, $\|H\| \in L^3(|\,\mathrm{dvol}\,|)$. Finally, given $\epsilon > 0$, there exists Σ_ϵ of this form so $\|H\| \notin L^{3-\epsilon}(|\,\mathrm{dvol}\,|)$. Consequently, $p = 3$ is the optimal universal index.*

If there are complex roots of $\mathcal{P}_1$ or of $\mathcal{P}_2$ which are dominant, then $|K|[\Sigma]$ can be infinite. Therefore, the assumption that the real roots are dominant is necessary in Assertion 1.

It is possible to reduce the computation of $K[\Sigma]$ to a question on the associated curves σ_{P_1} and σ_{P_2} using the Gauss–Bonnet Theorem. Let σ be the curve defined by an ODE with simple roots and dominant real roots. Set

$$\Theta_\sigma(t) := \frac{(\dot\sigma(t) \wedge \sigma(t), \dot\sigma(t) \wedge \ddot\sigma(t))}{\|\dot\sigma(t) \wedge \sigma(t)\| \cdot \|\dot\sigma(t)\|^3}\,.$$

We apply the Cauchy–Schwarz inequality to see $|\Theta_\sigma(t)| \le \kappa_\sigma(t)$ and hence $\Theta_\sigma ds$ is integrable and we may set:

$$\Theta[\sigma] := \int_{-\infty}^{\infty} \Theta_\sigma ds = \int_{-\infty}^{\infty} \frac{(\dot\sigma(t) \wedge \sigma(t), \dot\sigma(t) \wedge \ddot\sigma(t))}{\|\dot\sigma(t) \wedge \sigma(t)\| \cdot \|\dot\sigma(t)\|^2} dt\,.$$

Lemma 4.3 Let $\gamma_{\pm r}(t) := \Sigma(t, \pm r) = \sigma_1(t) \otimes \sigma_2(\pm r)$. Let κ_g be the geodesic curvature defined by the inward pointing unit normal. If all the roots of $\mathcal{P}_1$ and of $\mathcal{P}_2$ are simple and if the real roots are dominant, then:

$$\lim_{r\to\infty} \int_{-r}^{r} \kappa_g(\gamma_{\pm r})(t) ds = -\Theta[\sigma_1]\,.$$

Proof. Let $\Sigma_{/i} := \partial_{t_i}\Sigma$ and $\Sigma_{/ij} := \partial_{t_i}\partial_{t_j}\Sigma$. We will use the inward unit normal to apply the Gauss–Bonnet Theorem. This points in the direction of $\mp\{\Sigma_{/1} \wedge \Sigma_{/2}(t, \pm r)\}$. One has:

$$\kappa_g(t, \pm r) ds = \mp\{(\Sigma_{/1} \wedge \Sigma_{/2}, \Sigma_{/1} \wedge \Sigma_{/11}) \cdot g^{-1} \|\Sigma_{/1}\|^{-2}\}(t, \pm r) dt\,.$$

First let $t_2 = r$. Decompose $\sigma_1(t_1) = \sum_i e^{r_i t_1} e_i$ and $\sigma_2(t_2) = \sum_j e^{s_j t_2} f_j$ relative to bases $\{e_i\}$ for $\mathbb{R}^{n_1}$ and $\{f_j\}$ for $\mathbb{R}^{n_2}$. We express $\sigma_2(t_2) = e^{s_1 t_2}(f_1 + \mathcal{E}(t_2))$ where the remainder $\mathcal{E}(t_2)$ is exponentially suppressed, i.e., satisfies an estimate of the form $\|\mathcal{E}(t_2)\| \le e^{-\epsilon t_2}$ for some $\epsilon > 0$ if $t_2 >> 0$. In this setting, we shall write $\sigma_2(t_2) \sim e^{s_1 t_2} f_1$. We compute:

$$\begin{aligned}
&\Sigma_{/1} \sim \dot\sigma_1 \otimes e^{s_1 r} f_1, && \Sigma_{/2} \sim \sigma_1 \otimes s_1 e^{s_1 r} f_1,\\
&g = \|\Sigma_{/1} \wedge \Sigma_{/2}\| \sim |s_1| e^{2 s_1 r} \|\dot\sigma_1 \wedge \sigma_1\|, && \Sigma_{/11} \sim \ddot\sigma_1 \otimes e^{s_1 r} f_1,\\
&\kappa_g(\gamma_r) ds = -(\Sigma_{/1} \wedge \Sigma_{/2}, \Sigma_{/1} \wedge \Sigma_{/11}) g^{-1} \|\Sigma_{/1}\|^{-2} dt\\
&\qquad \sim -\frac{s_1 e^{4 s_1 r}}{|s_1| e^{4 s_1 r}} \frac{(\dot\sigma_1(t_1) \wedge \sigma_1(t_1), \dot\sigma_1(t_1) \wedge \ddot\sigma_1(t_1))}{\|\sigma_1(t_1) \wedge \dot\sigma_1(t_1)\| \cdot \|\dot\sigma_1(t_1)\|^2} dt\,.
\end{aligned}$$

This gives $-\Theta(\sigma_1) dt$ in the limit since $s_1 > 0$. We do not need to change the sign of the normal but again get a negative sign if $s_k < 0$ since $+\frac{s_k}{|s_k|} = -1$. □

Theorem 4.4 *If the roots of $\mathcal{P}_i$ are simple and if the real roots are dominant for $i = 1, 2$, then:*

1. $0 = K[\Sigma] - 2\Theta[\sigma_1] - 2\Theta[\sigma_2] + 2\pi$.
2. *If there are no complex roots, then* $|K[\Sigma]| \le \frac{\pi}{2}\{n_1(n_1 - 1) + n_2(n_2 - 1)\} + 2\pi$.
3. *If P_1 and P_2 are second order ODEs, then* $K[\Sigma] = 0$.

Proof. We apply the Gauss–Bonnet Theorem to the square $\Sigma([-r, r] \times [-r, r])$. Let α_i be the interior angles. We then have:

$$\begin{aligned}
2\pi &= \int_{-r}^{r}\int_{-r}^{r} K(t_1, t_2) g\, dt_1 dt_2 + \sum_{i=1}^{4} (\pi - \alpha_i) + \int_{-r}^{r} \kappa_g(\Sigma(t, r)) ds\\
&\quad + \int_{-r}^{r} \kappa_g(\Sigma(t, -r)) ds + \int_{-r}^{r} \kappa_g(\Sigma(r, t)) ds + \int_{-r}^{r} \kappa_g(\Sigma(-r, t)) ds\,.
\end{aligned}$$

Let α_1 be the angle at $\Sigma(r, r)$. Since $\Sigma_{/1}(r, r) \sim r_1 \Sigma(r, r)$ and since $\Sigma_{/2}(r, r) \sim s_1 \Sigma(r, r)$, $\Sigma_{/1}$ and $\Sigma_{/2}$ point in approximately the same direction. Consequently, $\cos(\alpha_1) \sim 1$ and $\alpha_1 \sim 0$. Keeping careful track of the signs shows the other angles also are close to 0. Assertion 1 then follows from Lemma 4.3. Since $|\Theta[\sigma_i]| \le \kappa[\sigma_i]$, Assertion 2 follows from Assertion 1 and from Assertion 2 of Theorem 4.1. To prove Assertion 3, we make a direct computation. Let $\sigma(t) := (e^{at}, e^{bt})$. Then

$$\Theta[\sigma] = \int_{-\infty}^{\infty} \frac{|(a - b)ab| e^{(a+b)t}}{a^2 e^{2at} + b^2 e^{2bt}} dt = \int_{-\infty}^{\infty} \frac{|(a - b)ab| e^{(a-b)t}}{a^2 e^{2(a-b)t} + b^2} dt\,.$$

We have $a - b > 0$. We change variables setting $x := e^{(a-b)t}$ to express

$$\Theta[\sigma] = \int_0^{\infty} \frac{|ab|}{a^2 x^2 + b^2} dx = \int_0^{\infty} \frac{|a|}{|b|} \frac{1}{\frac{a^2}{b^2} x^2 + 1} dx\,.$$

We again change variables setting $y = \frac{|a|}{|b|}x$ to express

$$\Theta[\sigma] = \int_0^\infty \frac{1}{y^2+1} dy = \frac{\pi}{2}.$$

Assertion 3 now follows from Assertion 1. □

4.1.1 EXAMPLE. Let $\sigma_1(t_1) = (e^{2t_1}, e^{-3t_1})$ and $\sigma_2(t_2) = (e^{2t_2}, e^{-3t_2})$. We may compute:

$$gK = \frac{125e^{4(t_1+t_2)}(-9+4e^{10(t_1+t_2)})}{9e^{10t_1}+9e^{10t_2}+e^{10(t_1+t_2)}+4e^{10(2t_1+t_2)}+4e^{10(t_1+2t_2)}}.$$

Let D be the denominator. Then

$$\begin{array}{lll} D \geq e^{10(2t_1+t_2)} & \text{and} \quad D \geq e^{10(t_1+2t_2)} & \text{so} \\ D^2 \geq e^{10(3t_1+3t_2)} & \text{and} \quad D \geq e^{15t_1+15t_2}. & \end{array}$$

The numerator is bounded by $e^{14(t_1+t_2)}$. Thus, gK is integrable. Let $\xi(t) := e^{10(t_1+t_2)}$. The Gaussian curvature changes sign. It is positive for $\xi \geq \frac{9}{4}$ and negative for $\xi(t_1, t_2) < \frac{9}{4}$. It does not vanish identically and Assertion 2 of Theorem 4.4 is non-trivial.

If we set $\sigma_1(t_1) = (e^{2t_1}, 1, e^{-3t_1})$ and $\sigma_2(t_2) = (e^{2t_2}, e^{-3t_2})$, then

$$\int_\Sigma gK \,\text{dvol} \approx -.951333 \quad \text{and} \quad \int_\Sigma |gK| \,\text{dvol} \approx 1.09409.$$

Thus, the total Gaussian curvature is non-zero for this example and again K changes sign.

4.2 VOLUME OF GEODESIC BALLS

We recall some notation established previously in Book I. Let g be a pseudo-Riemannian metric on a smooth manifold M of dimension m; g is a smooth section to $S^2(T^*M)$ and can be regarded as a smooth family of non-degenerate symmetric bilinear forms on each tangent space T_PM. Let ∇ be the Levi–Civita connection; it is characterized by the properties:

$$\nabla_X Y - \nabla_Y X = [X, Y] \quad \text{and} \quad X(g(Y,Z)) = g(\nabla_X Y, Z) + g(Y, \nabla_X Z).$$

This connection was first studied by the Italian mathematician Tullio Levi–Civita.

T. Levi–Civita (1873–1941)

Let $\mathcal{R}$ and $|\,\mathrm{dvol}\,|$ be the curvature operator and the Riemannian measure, respectively:

$$\mathcal{R}(X,Y) := \nabla_X\nabla_Y - \nabla_Y\nabla_X - \nabla_{[X,Y]} \quad \text{and} \quad |\,\mathrm{dvol}\,| = |\det(g_{ij})|^{\frac{1}{2}} dx^1 \cdot \cdots \cdot dx^m .$$

4.2.1 THE CURVATURE TENSOR IN GEODESIC COORDINATES. Let $\exp_P$ be the exponential map defined by g. It is a diffeomorphism from a neighborhood of 0 in T_PM to a neighborhood of P in M which is characterized by the property that the curve $\sigma_v : t \to \exp_P(tv)$ is a geodesic with $\sigma_v(0) = P$ and $\dot\sigma_v(0) = v$ for any $v \in T_PM$. Let $\log_P$ be the local inverse. If $\vec{e} := \{e_1, \dots, e_m\}$ is a basis for T_PM, let

$$\psi(\vec{x}) := \exp_P(x^1e_1 + \cdots + x^me_m) .$$

This gives a system of local coordinates on M called *geodesic coordinates*. We adopt the *Einstein convention* and sum over repeated indices where one index is up and one index is down and set, for example, $x^ie_i := x^1e_1 + \cdots + x^me_m$. Let i, j, k, ℓ be indices with $1 \le i, j, k, \ell \le m$. Let:

$$\begin{aligned} g_{ij} &:= g(\partial_{x^i}, \partial_{x^j}), & g_{ij/k} &:= \partial_{x^k} g_{ij}, \\ g_{ij/k\ell} &:= \partial_{x^\ell}\partial_{x^k} g_{ij}, & R_{ijk\ell} &:= g(\mathcal{R}(\partial_{x^i}, \partial_{x^j})\partial_{x^k}, \partial_{x^\ell}) . \end{aligned}$$

In a system of geodesic coordinates, we can express the components of the curvature tensor in terms of the 2-jets of the metric, and we can express the 2-jets of the metric in terms of the components of the curvature tensor at the center of the coordinate system.

Lemma 4.5 Let P be a point of a pseudo-Riemannian manifold (M, g) and let $\vec{x}$ be a system of geodesic coordinates centered at P. We have:

$$\begin{aligned} & g_{ij/k}(P) = 0, \\ & g_{ik/j\ell}(P) = -(g_{kj/i\ell} + g_{k\ell/ij})(P) = g_{j\ell/ik}(P), \\ & R_{ijk\ell}(P) = g_{ik/j\ell}(P) - g_{i\ell/jk}(P), \\ & 3g_{ik/j\ell}(P) = R_{ijk\ell}(P) - R_{i\ell jk}(P) . \end{aligned}$$

Proof. If s, u, v, and w are vectors in $\mathbb{R}^m$, denote the corresponding (constant) coordinate vector fields by:

$$S(\cdot) = s^i\partial_{x^i}, \quad U(\cdot) = u^i\partial_{x^i}, \quad V(\cdot) = v^i\partial_{x^i}, \quad \text{and} \quad W(\cdot) = w^i\partial_{x^i} .$$

The bracket of any pair of these vector fields vanishes. We have

$$g(\nabla_V V, W) = Vg(V, W) - \tfrac{1}{2}Wg(V, V) .$$

The curve $t \to tv$ is a geodesic so $g(\nabla_V V, W)(tv) = 0$ for $t \in [0, \infty)$. Consequently,

$$0 = 2Vg(V, W)(P) - Wg(V, V)(P) \quad \text{and} \tag{4.2.a}$$
$$0 = 2VVg(V, W)(P) - VWg(V, V)(P)\,. \tag{4.2.b}$$

We set $V = W$ in Equation (4.2.a) to see $Vg(V, V)(P) = 0$. Let $V(\varepsilon) := V + \varepsilon W$. The identity $\partial_\epsilon \left\{V(\epsilon)g(V(\epsilon), V(\epsilon))(P)\right\}|_{\epsilon=0} = 0$ implies that

$$0 = Wg(V, V)(P) + 2Vg(V, W)(P)\,. \tag{4.2.c}$$

We use Equations (4.2.a) and (4.2.c) to see that

$$0 = Wg(V, V)(P)\,. \tag{4.2.d}$$

We polarize this identity. Let $V(\varepsilon) := V + \varepsilon U$. We differentiate the identity of Equation (4.2.d), and set $\varepsilon = 0$ to see

$$0 = Wg(U, V)(P) \quad \text{so} \quad g_{ij/k}(P) = 0\,.$$

We set $V = W$ in Equation (4.2.b) to see that $0 = VVg(V, V)(P)$. We polarize this identity. Let $V(\varepsilon) := V + \varepsilon W$. We differentiate the identity $0 = V(\epsilon)V(\epsilon)g(V(\epsilon), V(\epsilon))(P)$ with respect to ε and set $\varepsilon = 0$ to see $0 = VVg(V, W)(P) + VWg(V, V)(P)$. We use this identity and Equation (4.2.b) to see:

$$0 = VVg(V, W)(P) \quad \text{and} \quad 0 = VWg(V, V)(P)\,. \tag{4.2.e}$$

We polarize these identities. Let $V(\varepsilon) := V + \varepsilon U$. We differentiate the identities of Equation (4.2.e) with respect to ε and set $\varepsilon = 0$ to see that

$$\begin{aligned} 0 &= 2UVg(V, W)(P) + VVg(U, W)(P), \\ 0 &= UWg(V, V)(P) + 2VWg(U, V)(P)\,. \end{aligned} \tag{4.2.f}$$

Note that $VVg(U, W)$ is symmetric in U and W. We use the relations of Equation (4.2.f) twice to see that for all U, V, W:

$$VVg(U, W)(P) = -2UVg(V, W)(P) = -2WVg(U, V)(P) = UWg(V, V)(P)\,.$$

Let $V(\varepsilon) := V + \varepsilon S$. We differentiate this identity with respect to ε and set ε to 0 to see

$$\begin{aligned} &VSg(U, W)(P) = -USg(W, V)(P) - UVg(W, S)(P) = UWg(V, S)(P) \quad \text{so} \\ &g_{ik/j\ell}(P) = -(g_{kj/i\ell} + g_{k\ell/ij})(P) = g_{j\ell/ik}(P)\,. \end{aligned}$$

Because $g_{ab/c}(P) = 0$, we have that

$$R_{ijk\ell}(P) = \tfrac{1}{2}\{g_{j\ell/ik} + g_{ik/j\ell} - g_{jk/i\ell} - g_{i\ell/jk}\}(P) = g_{ik/j\ell}(P) - g_{i\ell/jk}(P)\,.$$

We complete the proof by using the identities we have already derived to see that

$$R_{ijk\ell}(P) - R_{i\ell jk}(P) = \{g_{ik/j\ell} - g_{i\ell/jk} - g_{ij/k\ell} + g_{ik/j\ell}\}(P) = 3g_{ik/j\ell}(P)\,. \qquad \square$$

4.2.2 THE GROWTH OF VOLUME OF BALLS. The scalar curvature is defined by setting $\tau := g^{jk} R_{ijk}{}^{i}$; it plays a crucial role in the Gauss–Bonnet Theorem. It is also closely related to the volume of small geodesic balls. Let P be a point of a Riemannian manifold of dimension m. Let $V_r^M(P)$ be the volume of the geodesic ball $B_r^M(P)$ of radius r about P. Let $V_r^{\mathbb{R}^m}$ be the volume of the metric ball of radius r about the origin in $\mathbb{R}^m$.

Theorem 4.6 *As* $r \downarrow 0$, *we have* $V_r^M = V_r^{\mathbb{R}^m}\left\{1 - r^2\frac{\tau}{6(m+2)} + O(r^4)\right\}$.

We remark that there is a full normalized Taylor series in r^2 and refer to Gray [23] for further details. If $\Delta\tau = -g^{ij}\tau_{;ij}$, then the next term in the expansion is:

$$r^4\frac{-3\|R\|^2 + 8\|\rho\|^2 + 5\tau^2 + 18\Delta\tau}{360(m+2)(m+4)}.$$

Proof. We work in geodesic polar coordinates. We expand

$$\begin{aligned}
g_{ij}(\vec{x}) &= \delta_{ij} + \tfrac{1}{2}g_{ij/k\ell}x^k x^\ell + O(\|\vec{x}\|^4),\\
\det(g_{ij})(\vec{x}) &= 1 + \frac{1}{2}\sum_{i,k,\ell} g_{ii/k\ell}x^k x^\ell + O(\|\vec{x}\|^4),\\
\det(g_{ij})(\vec{x})^{\frac{1}{2}} &= 1 + \frac{1}{4}\sum_{i,k,\ell} g_{ii/k\ell}x^k x^\ell + O(\|\vec{x}\|^4),\\
V_r^M &= \int_{\|\vec{x}\|^2\le r}\left\{1 + \frac{1}{4}\sum_{i,k,\ell} g_{ii/k\ell}x^k x^\ell + \cdots\right\} d\vec{x}.
\end{aligned}$$

The integral of the cubic term vanishes and $\int x^k x^\ell = 0$ for $k \neq \ell$. We compute:

$$\begin{aligned}
\int_{\|\vec{x}\|^2\le r} d\vec{x} &= \int_{\varrho=0}^{r}\int_{\theta\in S^{m-1}} \varrho^{m-1}|\operatorname{dvol}|(\theta)d\varrho = \operatorname{Vol}(S^{m-1})r^m/m,\\
\int_{\|\vec{x}\|^2\le r} x^1x^1 d\vec{x} &= \frac{1}{m}\int_{\|\vec{x}\|^2\le r}\|\vec{x}\|^2 d\vec{x} = \frac{1}{m}\int_{\varrho=0}^{r}\int_{\theta\in S^{m-1}} \varrho^{m-1}\varrho^2|\operatorname{dvol}|(\theta)d\varrho\\
&= \frac{1}{m(m+2)}\operatorname{Vol}(S^{m-1})r^{m+2},\\
V_r^M &= \frac{1}{m}r^m\operatorname{Vol}(S^{m-1})\left\{1 + r^2\frac{1}{4(m+2)}\sum_{i,j} g_{ii/jj} + \cdots\right\}\\
&= V_r^{\mathbb{R}^m}\left\{1 - r^2\frac{1}{6(m+2)}R_{ijji}\right\} + \cdots.
\end{aligned}$$

□

4.3 HOLOMORPHIC GEOMETRY

We say that a function $f(z) = f(x, y)$ from an open subset $\mathcal{O}$ of $\mathbb{C} = \mathbb{R}^2$ to $\mathbb{C} = \mathbb{R}^2$ is *holomorphic* if it is complex differentiable, i.e., if

$$f'(z) = \lim_{\delta \to 0} \frac{f(z + \delta) - f(z)}{\delta}$$

exists and is independent of the particular complex direction δ which is used to approach the origin for every $z \in \mathcal{O}$. Decompose $f = u + \sqrt{-1}v$ into real and imaginary parts. Letting δ approach 0 along the real or the complex axis then yields the *Cauchy–Riemann Equations* [13, 54]

$$\partial_x u = \partial_y v \quad \text{and} \quad \partial_x v = -\partial_y u\,. \tag{4.3.a}$$

Holomorphic functions have many properties; the sum, product, difference, quotient (if appropriate), and composition of holomorphic functions is holomorphic. The Taylor series of a holomorphic function converges uniformly to the holomorphic function near each point in the domain; conversely, since the uniform limit of holomorphic functions is again holomorphic, a complex-valued function of a complex variable which is given by a convergent Taylor series is necessarily holomorphic. A vector-valued function $F : \mathcal{O} \subset \mathbb{C}^n \to \mathbb{C}^p$ is said to be holomorphic on $\mathcal{O}$ if each of the components of F is holomorphic in each of the variables separately. The complex Jacobian is the matrix $F'_c := \partial_{z^\alpha} F_\beta$. If $n = p$ and if $\det(F'_c) \neq 0$, then F is a local diffeomorphism and a local inverse F^{-1} is again holomorphic. If $f : \mathcal{O} \subset \mathbb{C}^n \to \mathbb{C}$ is a scalar non-constant holomorphic function and if $\mathcal{O}$ is connected, then $\{z : f(z) \neq 0\}$ is again connected.

An *almost complex structure* on a smooth manifold M is a smooth endomorphism J of TM so that $J^2 = -\operatorname{Id}$. A coordinate system (x^α, y^α) is said to be a *holomorphic coordinate system* if $J\partial_{x^\alpha} = \partial_{y^\alpha}$ and $J\partial_{y^\alpha} = -\partial_{x^\alpha}$. The *Nijenhuis tensor* N_J [47] is given by:

$$N_J(X, Y) := [X, Y] + J[JX, Y] + J[X, JY] - [JX, JY]\,.$$

The *complex tangent bundle* is the eigen-subbundle of the complexified tangent bundle defined by setting:

$$T_{\mathbb{C}}(M) = \{v \in TM \otimes_{\mathbb{R}} \mathbb{C} : Jv = \sqrt{-1}v\}\,.$$

We extend the Lie bracket from real tangent bundle TM to the complexified tangent bundle $TM \otimes_{\mathbb{R}} \mathbb{C}$ to be complex bilinear. We say that $T_{\mathbb{C}}(M)$ is *closed under bracket* if given sections ϕ and ψ in $C^\infty T_{\mathbb{C}}(M)$, we have that the bracket $[\phi, \psi]$ belongs to $C^\infty T_{\mathbb{C}}(M)$ as well. We refer to Newlander and Nirenberg [46] for the proof of the following result which is called the *Newlander–Nirenberg* Theorem.

Louis Nirenberg (1925–)

The following is a deep result in the theory of partial differential equations which is beyond the scope of this book and which can be regarded as a complex version of the real Frobenius Theorem (see Theorem 2.8 in Book I).

Theorem 4.7 *Let (M, J) be an almost complex manifold of dimension $m = 2\bar{m}$. The following conditions are equivalent and if any is satisfied, then (M, J) is said to be a complex manifold and the structure J is said to be an integrable complex structure:*

1. *There are holomorphic coordinate charts covering M.*
2. *The* Nijenhuis tensor *is 0.*
3. *The complex tangent bundle $T_{\mathbb{C}}(M)$ is closed under bracket.*

4.3.1 ISOTHERMAL COORDINATES. Recall that a system of local coordinates (x, y) on a Riemann surface is said to be *isothermal* if $ds^2 = e^{2h}(dx^2 + dy^2)$ for some smooth function h.

Corollary 4.8 There exist isothermal coordinate charts covering any Riemann surface.

Proof. Let P be a point of a Riemann surface M. Let $\{e^1, e^2\}$ be a local orthonormal frame for TM which is defined near P. Then the canonical almost complex structure

$$Je^1 = e^2 \quad \text{and} \quad Je^2 = -e^1$$

is integrable since $T_{\mathbb{C}}M$ is 1-dimensional. Therefore, we can choose local holomorphic coordinates so $J\partial_x = \partial_y$ and $J\partial_y = -\partial_x$. Since J is unitary, $J^*g = g$. Consequently,

$$g(\partial_x, \partial_x) = g(\partial_y, \partial_y) \quad \text{and} \quad g(\partial_x, \partial_y) = 0$$

so $g = e^{2h}(dx^2 + dy^2)$ for some smooth conformal factor h. □

Let (M, J) be a complex manifold. The transition functions relating any two holomorphic coordinate systems are holomorphic. Conversely, if (M, J) is a manifold which admits a cover by charts such that the transition functions are holomorphic, then we can recover J by setting

$$J\partial_{x^\alpha} = \partial_{y^\alpha} \quad \text{and} \quad J\partial_{y^\alpha} = -\partial_{x^\alpha}\,.$$

Equation (4.3.a) then ensures that J is well-defined and independent of the particular local coordinate system chosen. Let $\mathcal{O}_\alpha$ and $\mathcal{O}_\beta$ be open subsets of $\mathbb{C}^{\bar{m}}$ and let $F : \mathcal{O}_\alpha \to \mathcal{O}_\beta$ be a holomorphic diffeomorphism. Let $F_c' := \partial_{z^\alpha} F_\beta$ be the holomorphic Jacobian and let F' be the ordinary Jacobian. The Cauchy–Riemann equations imply $\det(F') = |\det(F_c')|^2 > 0$. Consequently, every holomorphic manifold is orientable.

4.3.2 THE OPERATORS ∂ AND $\bar{\partial}$.

Let (M, J) be a complex manifold and let $(x^1, \ldots, x^{\bar{m}}, y^1, \ldots, y^{\bar{m}})$ be a system of local holomorphic coordinates. Introduce holomorphic variables $z^\alpha := x^\alpha + \sqrt{-1}y^\alpha$ and define

$$\begin{aligned} dz^\alpha &= dx^\alpha + \sqrt{-1}dy^\alpha, & d\bar{z}^\alpha &= dx^\alpha - \sqrt{-1}dy^\alpha, \\ \partial_{z^\alpha} &:= \tfrac{1}{2}(\partial_{x^\alpha} - \sqrt{-1}\partial_{y^\alpha}), & \partial_{\bar{z}^\alpha} &:= \tfrac{1}{2}(\partial_{x^\alpha} + \sqrt{-1}\partial_{y^\alpha}). \end{aligned}$$

Extend the usual pairing $\langle \cdot, \cdot \rangle$ between a cotangent and a tangent vector to be complex bilinear. We then have:

$$\langle dz^\alpha, \partial_{z^\beta} \rangle = \langle d\bar{z}^\alpha, \partial_{\bar{z}^\beta} \rangle = \delta^\alpha_\beta \quad \text{and} \quad \langle dz^\alpha, \partial_{\bar{z}^\beta} \rangle = \langle d\bar{z}^\alpha, \partial_{z^\beta} \rangle = 0 .$$

Let f be a smooth complex-valued function. We sum over repeated indices to define:

$$\partial f := (\partial_{z^\alpha} f) dz^\alpha \quad \text{and} \quad \bar{\partial} f := (\partial_{\bar{z}^\alpha} f) d\bar{z}^\alpha .$$

The Cauchy–Riemann equations show that a smooth complex-valued function f is holomorphic if and only if $\bar{\partial} f = 0$. If $I = \{1 \le i_1 < \cdots < i_p \le \bar{m}\}$ and $J = \{1 \le j_1 < \cdots \le j_q \le \bar{m}\}$ are collections of indices, set:

$$dz^I := dz^{i_1} \wedge \cdots \wedge dz^{i_p} \quad \text{and} \quad d\bar{z}^J := d\bar{z}^{j_1} \wedge \cdots \wedge d\bar{z}^{j_q} .$$

Let $\Lambda^{p,q} := \operatorname{span}_{|I|=p,|J|=q} dz^I \wedge d\bar{z}^J$. The Cauchy–Riemann equations show that ∂f, $\bar{\partial} f$, and $\Lambda^{p,q}$ are invariantly defined, i.e., independent of the particular holomorphic coordinate system which is chosen. We may decompose the complex exterior algebra in the form:

$$\Lambda^n(M) \otimes_{\mathbb{R}} \mathbb{C} = \oplus_{p+q=n} \Lambda^{p,q} M .$$

We extend ∂ and $\bar{\partial}$ to maps

$$\begin{aligned} \partial(f_{I,J} dz^I \wedge d\bar{z}^J) &= \partial f_{I,J} \wedge dz^I \wedge d\bar{z}^J : C^\infty(\Lambda^{p,q} M) \to C^\infty(\Lambda^{p+1,q} M) \quad \text{and} \\ \bar{\partial}(f_{I,J} dz^I \wedge d\bar{z}^J) &= \bar{\partial} f_{I,J} \wedge dz^I \wedge d\bar{z}^J : C^\infty(\Lambda^{p,q} M) \to C^\infty(\Lambda^{p,q+1} M) . \end{aligned}$$

Again, these are independent of the particular holomorphic coordinate system chosen. We have:

$$d = \partial + \bar{\partial}, \qquad \partial\partial = 0, \qquad \bar{\partial}\bar{\partial} = 0, \qquad \partial\bar{\partial} + \bar{\partial}\partial = 0 .$$

4.3.3 COMPLEX PROJECTIVE SPACE. We continue the discussion of Section 2.3.6 of Book I. Let

$$\mathfrak{CP}^m := \{\rho \in M_{m+1}(\mathbb{C}) : \rho^* = \rho, \quad \rho^2 = \rho, \quad \operatorname{Rank}(\rho) = 1\}$$

be the set of orthogonal projections of rank 1 in $M_{m+1}(\mathbb{C})$; $\mathfrak{CP}^m$ is a compact metric space. Let $\mathbb{CP}^m$ be the set of complex lines in $\mathbb{C}^{m+1}$ through the origin, i.e., 1-dimensional subspaces of $\mathbb{C}^{m+1}$. If $\xi \in \mathbb{CP}^m$, let ρ_ξ be orthogonal projection on ξ; this identifies $\mathbb{CP}^m$ with $\mathfrak{CP}^m$ and gives $\mathbb{CP}^m$ the structure of a compact metric space. The non-zero complex numbers $\mathbb{C} - \{0\}$ act on $\mathbb{C}^{m+1} - \{0\}$ by scalar multiplication; this action restricts to an action of S^1 on S^{2m+1}. The map $\pi : z \to z \cdot \mathbb{C}$ defines a homeomorphism from S^{2m+1}/S^1 with the quotient topology to $\mathbb{CP}^m$ or from $\{\mathbb{C}^{m+1} - \{0\}\}/\{\mathbb{C} - \{0\}\}$ with the quotient topology to $\mathbb{CP}^m$. We give $\mathbb{CP}^m$ the structure of a holomorphic manifold as follows. Let

$$U_i := \{z \in \mathbb{C}^{m+1} : z^i \neq 0\} \quad \text{and} \quad \{\mathcal{O}_i := \pi(U_i)\}$$

be the associated open covers of $\mathbb{C}^{m+1} - \{0\}$ and $\mathbb{CP}^m$, respectively. Let

$$F_i(w^1, \ldots, w^m) = (w^1, \ldots, 1, \ldots, w^m) : \mathbb{C}^m \to U_i$$

be defined by putting a 1 in the ith position. Then $\pi \circ F_i$ is a homeomorphism from $\mathbb{C}^m$ onto $\mathcal{O}_i$ that defines a local coordinate chart. More specifically, if $z \in \mathcal{O}_i$, let $z_i^j(z) := z^j / z^i$. Since $z_i^j(\lambda z) = z_i^j(z)$ for $\lambda \neq 0$, the z_i^j descend to define continuous functions on $\mathcal{O}_i$ which give local coordinates (called *homogeneous coordinates*) with

$$z_i^j(\pi F_i w) = \left\{ \begin{array}{ll} w^j & \text{for} \quad j < i \\ w^{j-1} & \text{for} \quad j > i \end{array} \right\} .$$

Since $z_i^j = z_k^j / z_k^i$, the transition functions are holomorphic so $\mathbb{CP}^m$ has the structure of a *complex manifold* and $\pi \circ F_i$ is a holomorphic diffeomorphism from $\mathbb{C}^m$ to $\mathcal{O}_i$. A point of $\mathbb{CP}^m$ is a line in $\mathbb{C}^{m+1}$. The fiber of the *tautological line bundle* $\mathbb{L}$ is defined to be that line, i.e.,

$$\mathbb{L} := \{\xi \times z \in \mathbb{CP}^m \times \mathbb{C}^{m+1} : z \in \xi\} . \tag{4.3.b}$$

We define local sections to $\mathbb{L}$ over the open charts $\mathcal{O}_i$ by setting:

$$s_i(\xi) := (\xi, z_i^1(\xi), \ldots, z_i^{m+1}(\xi)) \quad \text{for} \quad \xi \in \mathcal{O}_i .$$

Since $s_i(\xi) = z_i^j(\xi) s_j(\xi)$ on $\mathcal{O}_i \cap \mathcal{O}_j$, $\mathbb{L}$ is a holomorphic line bundle over $\mathbb{CP}^m$.

If $m = 1$, then we have two coordinate systems $(\mathbb{C}, z_1)$ and $(\mathbb{C}, z_2)$ and the transition rule is $z = w^{-1}$ on $\mathbb{C} - \{0\}$. This is the Riemann sphere so $\mathbb{CP}^1 = S^2$. This can also be seen combinatorially. Let (w_1, w_2) be the usual coordinates on $\mathbb{C}^2$. Define

$$F(w_1, w_2) := (2\Re(w_1 \bar{w}_2), 2\Im(w_1 \bar{w}_2), |w_1|^2 - |w_2|^2) \in \mathbb{R}^3 .$$

Let $(w_1, w_2) \in S^3$. Then:

$$\begin{aligned}\|F(w_1, w_2)\|^2 &= 4\Re(w_1\bar{w}_2)^2 + 4\Im(w_1\bar{w}_2)^2 + |w_1|^4 + |w_2|^4 - 2|w_1|^2|w_2|^2 \\ &= 4|w_1\bar{w}_2|^2 + |w_1|^4 + |w_2|^4 - 2|w_1|^2|w_2|^2 \\ &= (|w_1|^2 + |w_2|^2)^2 = 1\,.\end{aligned}$$

Thus, F takes values in the unit sphere S^2. Since $F(\lambda w_1, \lambda w_2) = F(w_1, w_2)$, F defines a smooth map from $\mathbb{CP}^1$ to S^2. It is easily verified that the map is bijective and that the Jacobian is non-singular. Therefore, F provides the requisite identification of $\mathbb{CP}^1$ with S^2.

4.4 KÄHLER GEOMETRY

The German mathematician Erich Kähler introduced many of the concepts that we shall discuss here.

Erich Kähler (1906–2000)

If J is an almost complex structure on M, then a Riemannian metric g on M is said to be a *Hermitian metric* if $J^*g = g$. If g is an arbitrary Riemannian metric, we can always average g over the action of J to obtain a Hermitian metric $\tilde{g}$ by setting $\tilde{g} = (g + J^*g)/2$. The *Kähler form* Ω is defined by setting:

$$\Omega(X, Y) := g(X, JY)\,.$$

Lemma 3.13 in Book I generalizes to this context as the following result.

Theorem 4.9 Let P be a point of a Hermitian manifold (M, g, J).

1. The following conditions are equivalent and if any is satisfied, then (M, g, J) is said to be Kähler at P. If (M, g, J) is Kähler at every point, then (M, g, J) is said to be a Kähler manifold or simply Kähler.

    (a) $d\Omega(P) = 0$.

    (b) There exist local holomorphic coordinates $(w^1, \dots, w^{\bar{m}})$ for M centered at P so that $g = g(P) + O(|w|^2)$.

    (c) $(\nabla J)(P) = 0$.

    (d) $\nabla\Omega(P) = 0$.

(e) $d\Omega(P) = 0$ and $\delta\Omega(P) = 0$.

2. If (M, g, J) is Kähler, $J\mathcal{R}(x, y) = \mathcal{R}(x, y)J$ and $R(Jx, Jy, z, w) = R(x, y, z, w)$.

Proof. Extend g to $T_{\mathbb{C}}(M)$ to be complex bilinear in each factor. Let $w^a = x^a + \sqrt{-1}y^a$ be a system of local holomorphic coordinates on M where we let $1 \le a \le \bar{m}$. We then have

$$\partial_{w^a} := \tfrac{1}{2}(\partial_{x^a} - \sqrt{-1}\partial_{y^a}) \quad \text{and} \quad \partial_{\bar{w}^a} := \tfrac{1}{2}(\partial_{x^a} + \sqrt{-1}\partial_{y^a}).$$

We set $g^w_{a\bar{b}} := g(\partial_{w^a}, \partial_{\bar{w}^b})$; note that $g^w_{a\bar{b}} = \bar{g}^w_{b\bar{a}}$. We have $\Omega = \frac{1}{2\sqrt{-1}} g^w_{b\bar{d}} dw^b \wedge d\bar{w}^d$ so:

$$\begin{aligned} d\Omega &= \tfrac{1}{2\sqrt{-1}} \textstyle\sum_{b<c,d} (g^w_{c\bar{d}/b} - g^w_{b\bar{d}/c}) dw^b \wedge dw^c \wedge d\bar{w}^d \\ &\quad - \tfrac{1}{2\sqrt{-1}} \textstyle\sum_{b,c<d} (g^w_{b\bar{d}/\bar{c}} - g^w_{b\bar{c}/\bar{d}}) dw^b \wedge d\bar{w}^c \wedge d\bar{w}^d . \end{aligned}$$

Consequently, the condition $d\Omega(P) = 0$ implies that:

$$g^w_{b\bar{d}/c}(P) = g^w_{c\bar{d}/b}(P). \tag{4.4.a}$$

Let $\varepsilon_{a\bar{b}} := g^w_{a\bar{b}}(P)$. Make a quadratic change of coordinates to set $z^a := w^a + \xi^a{}_{bc} w^b w^c$ where the complex constants $\xi^a{}_{bc} = \xi^a{}_{cb}$ remain to be determined. Express

$$\begin{aligned} dz^a &= dw^a + 2\xi^a{}_{bc} w^b dw^c, \\ \partial_{w^c} &= \partial_{z^c} + 2\xi^a{}_{bc} w^b \partial_{z^a}, \\ g^w_{c\bar{d}} &= g^z_{c\bar{d}} + 2\varepsilon_{a\bar{d}} \xi^a{}_{bc} w^b + 2\varepsilon_{c\bar{a}} \bar{\xi}^a{}_{bd} \bar{w}^b + O(|w|^2), \\ g^w_{c\bar{d}/b}(P) &= g^z_{c\bar{d}/b}(P) + 2g_{a\bar{d}}(P)\xi^a{}_{bc} + O(|w|^2). \end{aligned}$$

We set $\xi^a{}_{bc} := \frac{1}{2} g^{a\bar{d}}(P) g^w_{c\bar{d}/b}(P)$; this is symmetric in $\{b, c\}$ by Equation (4.4.a) and defines an admissible change of coordinates with $g^z_{c\bar{d}/b}(P) = 0$. Taking the complex conjugate yields $g^z_{d\bar{c}/\bar{b}}(P) = 0$ as well. This shows that $dg^z(P) = 0$. Consequently, if Assertion 1-a is true, then Assertion 1-b is true. Assume Assertion 1-b holds. In a holomorphic coordinate system, ∇J is a linear expression in the first derivatives of the metric with coefficients which depend smoothly on the metric; such expressions vanish in a coordinate system as in Assertion 1-b. Therefore, Assertion 1-b implies Assertion 1-c. Assume Assertion 1-c holds. Since $\nabla\Omega$ can be computed in terms of ∇g (which is zero) and ∇J (which is zero), $\nabla\Omega = 0$ which verifies Assertion 1-d. Assume Assertion 1-d holds. We shall show presently in Theorem 5.14 that

$$d = \operatorname{ext} \circ \nabla \qquad \text{and} \qquad \delta = -\operatorname{int} \circ \nabla$$

where ext denotes exterior multiplication and int denotes the dual, interior multiplication. Consequently, $\nabla\Omega = 0$ implies $d\Omega = 0$ and $\delta\Omega = 0$ which establishes Assertion 1-e. It is clear that Assertion 1-e implies Assertion 1-a. This completes the proof of Assertion 1.

Suppose (M, g, J) is Kähler. Because $\nabla J = 0$, $\nabla_x J = J\nabla_x$ for any tangent vector x. Consequently, $\mathcal{R}(x, y)J = J\mathcal{R}(x, y)$. Furthermore,

$$
\begin{aligned}
R(Jx, Jy, z, w) &= R(z, w, Jx, Jy) = \langle \mathcal{R}(z, w)Jx, Jy\rangle \\
&= \langle J\mathcal{R}(z, w)x, Jy\rangle = \langle \mathcal{R}(z, w)x, y\rangle = R(z, w, x, y) = R(x, y, z, w)\,. \qquad \square
\end{aligned}
$$

Let (M, g, J) be a Kähler manifold. If $f \in C^\infty(M)$, we may define a real Hermitian symmetric bilinear form h_f and a corresponding real antisymmetric 2-form Ω_{h_f} by setting:

$$
h_f = \frac{\partial^2 f}{\partial_{z^\alpha} \partial_{\bar{z}^\beta}} dz^\alpha \circ d\bar{z}^\beta \quad \text{and} \quad \Omega_{h_f} = -\sqrt{-1}\partial\bar{\partial} f = -\sqrt{-1}\frac{\partial^2 f}{\partial_{z^\alpha} \partial_{\bar{z}^\beta}} dz^\alpha \wedge d\bar{z}^\beta\,.
$$

We then have $d\Omega_{h_f} = 0$ and, consequently, for small ε, $g + \varepsilon h_f$ is positive definite. Consequently, $g + \varepsilon h_f$ is a Kähler metric. This shows that if (M, g) admits one Kähler metric, then it admits many Kähler metrics.

The Kähler condition is much stronger than complex integrability. Let (M, g, J) be a compact Kähler manifold with a positive definite metric. As $\nabla\Omega = 0$, Ω is a harmonic form of degree two. Since $\Omega^{\bar{m}}$ is a multiple of the volume form, the Hodge Decomposition Theorem (see Theorem 5.13) shows that associated element $x = [\Omega]$ in de Rham cohomology satisfies $x^{\bar{m}} \neq 0$. This has topological implications and shows that not every manifold M admits a Kähler metric.

4.4.1 EXAMPLE. Let $M := \mathbb{C}^2 - \{0\}$ with the usual complex structure. Let $\lambda > 1$. We define an action τ of $\mathbb{Z}$ on M by setting $\tau_\lambda(n)(z_1, z_2) := \lambda^n(z_1, z_2)$. The resulting quotient manifold $M/\mathbb{Z}$ is then diffeomorphic to $S^1 \times S^3$. This shows that $S^1 \times S^3$ admits an integrable complex structure. On the other hand, as the second cohomology group $H^2(S^1 \times S^3) = 0$, $S^1 \times S^3$ admits no Kähler metric.

CHAPTER 5

de Rham Cohomology

In Chapter 5 we will explore de Rham cohomology. Since $d^2 = 0$, we may define:

$$H^p_{\mathrm{dR}}(M) := \frac{\ker\{d : C^\infty(\Lambda^p M) \to C^\infty(\Lambda^{p+1} M)\}}{\mathrm{range}\{d : C^\infty(\Lambda^{p-1} M) \to C^\infty(\Lambda^p M)\}} .$$

In Section 5.1, we discuss some basic properties of de Rham cohomology. Clifford algebras are at the heart of many of our computations so in Section 5.2, we present a brief introduction to the subject and refer to Atiyah, Bott, and Shapiro [3] for further details. In Section 5.3, we present the Hodge Decomposition Theorem which relates the spectrum of the Laplacian to de Rham cohomology. This leads to a discussion of Poincaré duality and the Künneth formula. Characteristic classes are introduced in Section 5.4 and the cohomology ring structure of complex projective space is determined.

5.1 BASIC PROPERTIES OF DE RHAM COHOMOLOGY

The Swiss mathematician Georges de Rham introduced many of the concepts we will discuss in this section.

Georges de Rham (1903–1990)

If $\omega_p \in C^\infty(\Lambda^p M)$ satisfies $d\omega_p = 0$, then ω_p is said to be a *closed differential form* and we let $[\omega_p] \in H^p_{\mathrm{dR}}(M)$ denote the associated element in de Rham cohomology. Let $\wedge$ denote the *wedge product* of differential forms defined in Section 2.3.1 of Book I. If $F : M \to N$ is a smooth map, let $F^* : C^\infty(\Lambda^p N) \to C^\infty(\Lambda^p M)$ be the *pullback* which was defined in Section 2.2.6 of Book I. De Rham cohomology satisfies the Eilenberg–Steenrod axioms [17, 18] (see the discussion in Chapter 8); these are properties that all homology and cohomology theories have. We shall replace the excision property by the equivalent Mayer–Vietoris sequence as it is more convenient for our purposes. De Rham cohomology is a contravariant functor from the category of smooth manifolds to the category of graded skew-commutative rings. We summarize the functorial properties of de Rham cohomology that we shall need in the following two theorems. The properties of Theorem 5.1 are immediate consequences of the definitions that we have given. The two properties of Theorem 5.2 are more difficult to establish and we shall postpone the proof of Theorem 5.2 until Section 8.2 as it relies upon some results in homological algebra that we shall derive subsequently. We shall let $[\omega_p]$ be the cohomology class of a closed smooth p-form ω_p, $[\tilde{\omega}_q]$ be the cohomology class of a closed smooth q-form $\tilde{\omega}_q$, and so forth. Recall that two points of M are said to be in the same *arc-component* if there is a continuous path joining them.

Theorem 5.1

1. *If M is a smooth manifold of dimension m, then $H^p_{\mathrm{dR}}(M) = 0$ for $p > m$.*
2. *If M is a single point, then $H^0_{\mathrm{dR}}(M) = [1] \cdot \mathbb{R}$.*
3. *If M has exactly ℓ arc-components, then $\dim\{H^0_{\mathrm{dR}}(M)\} = \ell$.*
4. *If M is the disjoint union of two open sets $\mathcal{O}_1$ and $\mathcal{O}_2$, then*
$$H^p_{\mathrm{dR}}(M) = H^p_{\mathrm{dR}}(\mathcal{O}_1) \oplus H^p_{\mathrm{dR}}(\mathcal{O}_2) \quad \textit{for any } p.$$
5. *Let ω_p (resp. $\tilde{\omega}_q$) be a closed p-form (resp. q-form). Then $[\omega_p] \wedge [\tilde{\omega}_q] := [\omega_p \wedge \tilde{\omega}_q]$ is well-defined in de Rham cohomology. This product is associative and one has:*
$$\begin{aligned}
&[\omega_p] \wedge [\tilde{\omega}_q] = (-1)^{pq}[\tilde{\omega}_q] \wedge [\omega_p],\\
&[1] \wedge [\omega_p] = [\omega_p] \wedge [1] = [\omega_p]\,.
\end{aligned}$$
6. *If $F : M \to N$, then $F^*[\omega_p] := [F^*\omega_p]$ is well-defined in de Rham cohomology. One has the identities $\mathrm{Id}^* = \mathrm{Id}$ and $(F \circ G)^* = G^* \circ F^*$.*

Proof. Assertion 1 and Assertion 2 are immediate since

$$\Lambda^p(M) = 0 \quad \text{for } p > m \quad \text{and} \quad \Lambda^0(\mathrm{pt}) = \mathbb{R}\,.$$

Let $f \in C^\infty(M)$. Then $df = 0$ implies that f is locally constant. The vector space of locally constant smooth functions on M can be identified with the real vector space generated by the arc-components. Assertion 3 now follows. Assertion 4 follows since

$$C^\infty(\Lambda^p M) = C^\infty(\Lambda^p \mathcal{O}_1) \oplus C^\infty(\Lambda^p \mathcal{O}_2)$$

and since the exterior derivative decouples, i.e., $d_M = d_{\mathcal{O}_1} \oplus d_{\mathcal{O}_2}$ if M is the disjoint union of $\mathcal{O}_1$ and $\mathcal{O}_2$. We now prove Assertion 5. Let $\omega_p \in C^\infty(\Lambda^p M)$ and let $\tilde{\omega}_q \in C^\infty(\Lambda^q M)$. By Lemma 2.15 in Book I,

$$d(\omega_p \wedge \tilde{\omega}_q) = d\omega_p \wedge \tilde{\omega}_q + (-1)^p \omega_p \wedge d\tilde{\omega}_q\,. \tag{5.1.a}$$

Suppose that $d\omega_p = 0$ and that $d\tilde{\omega}_q = 0$. By Equation (5.1.a), $d(\omega_p \wedge \tilde{\omega}_q) = 0$ so $[\omega_p \wedge \tilde{\omega}_q]$ is well-defined. If $\omega_p = d\phi_{p-1}$ for $\phi_{p-1} \in C^\infty(\Lambda^{p-1}M)$, then

$$d(\phi_{p-1} \wedge \tilde{\omega}_q) = \omega_p \wedge \tilde{\omega}_q .$$

Similarly, if $\tilde{\omega}_q = d\tilde{\phi}_{q-1}$ for $\tilde{\phi}_{q-1} \in C^\infty(\Lambda^{q-1}M)$, then

$$d((-1)^q \omega_p \wedge \tilde{\phi}_{q-1}) = \omega_p \wedge \tilde{\omega}_q .$$

It is now immediate that wedge product extends to de Rham cohomology and Assertion 5 follows from this observation and the analogous properties for differential forms. Since $dF^* = F^*d$, Assertion 6 follows. □

We now introduce some additional notational conventions.

5.1.1 NOTIONS FROM ALGEBRAIC TOPOLOGY.

1. We say that $f_i : M \to N$, $i = 0, 1$, are *homotopic maps* if there is a smooth map F from $M \times [0,1]$ to N so that $f_0(x) = F(x, 0)$ and $f_1(x) = F(x, 1)$. Note that f_0 is homotopic to f_1 and g_0 is homotopic to g_1 implies $f_0 \circ g_0$ is homotopic to $f_1 \circ g_1$. We say that M and N are *homotopy equivalent spaces* if there are maps $f : M \to N$ and $g : N \to M$ so that $f \circ g$ is homotopic to the identity on N and so that $g \circ f$ is homotopic to the identity on M. These are equivalence relations.

2. We say that $M_1 \subset M_2$ is a *deformation retract* of M_2 if there exists a smooth map F from $M_2 \times [0,1]$ to M_2 so that:

   (a) $F(x, t) = x$ for all $t \in [0, 1]$ and for any $x \in M_1$.

   (b) $F(y, 0) = y$ for any $y \in M_2$.

   (c) $F(y, 1) \in M_1$ for any $y \in M_2$.

   Let i be the inclusion of M_1 in M_2 and let $r(y) := F(y, 1)$ define a map from M_2 to M_1. We have $r \circ i = \mathrm{Id}_{M_1}$ and F provides a homotopy between $i \circ r$ and Id_{M_2}. Consequently, r and i are homotopy equivalences between M_1 and M_2.

3. We say that M is *contractible* if there exists a point P of M which is a deformation retract of M, i.e., if there exists a smooth map $F : M \times [0,1] \to M$ so that $F(P, t) = P$ for all t, if $F(x, 0) = x$ for all $x \in M$, and if $F(x, 1) = P$ for all $x \in M$. This implies that M is homotopy equivalent to the singleton set $\{P\}$.

4. Let A_n be a sequence of vector spaces for $n \geq 0$ and let $\alpha_n : A_n \to A_{n+1}$ be linear maps. We say

   $$0 \to A_0 \xrightarrow{\alpha_0} A_1 \xrightarrow{\alpha_1} A_2 \cdots$$

   is a *long exact sequence* if for all n we have $\ker\{\alpha_{n+1}\} = \mathrm{range}\{\alpha_n\}$.

We postpone the proof of the following results until Chapter 8 since the proof relies heavily on the machinery of homological algebra; the Mayer–Vietoris sequence was established by the German mathematician Walther Mayer and the Austrian mathematician Leopold Vietoris.

L. Vietoris (1891–2002)

Theorem 5.2

1. *If $\mathcal{O}_i$ are open subsets of M with $M = \mathcal{O}_1 \cup \mathcal{O}_2$, then there is a natural long exact sequence (called the Mayer–Vietoris sequence [41, 59]):*

$$\cdots H_{\mathrm{dR}}^{p-1}(M) \stackrel{i_1^*\oplus i_2^*}{\longrightarrow} H^{p-1}(\mathcal{O}_1) \oplus H^{p-1}(\mathcal{O}_2) \stackrel{j_1^*-j_2^*}{\longrightarrow} H_{\mathrm{dR}}^{p-1}(\mathcal{O}_1 \cap \mathcal{O}_2) \stackrel{\mathfrak{d}}{\longrightarrow} H_{\mathrm{dR}}^{p}(M) \cdots,$$

where $i_1 : \mathcal{O}_1 \to M$, $i_2 : \mathcal{O}_2 \to M$, $j_1 : \mathcal{O}_1 \cap \mathcal{O}_2 \to \mathcal{O}_1$, and $j_2 : \mathcal{O}_1 \cap \mathcal{O}_2 \to \mathcal{O}_2$ are the natural inclusions. The map $\mathfrak{d}$ is called the connecting homomorphism. If f is a smooth map from M to N, and if $\mathcal{U}_i := f^{-1}\mathcal{O}_i$ is the associated open cover of N, then $\mathfrak{d}_N f^* = f^*\mathfrak{d}_M$, i.e., $\mathfrak{d}$ is natural in this category.

2. *If $F_i : M \to N$ are homotopic smooth maps, then $F_0^* = F_1^* : H_{\mathrm{dR}}^p(N) \to H_{\mathrm{dR}}^p(M)$. In particular, if M and N are homotopy equivalent, then $H_{\mathrm{dR}}^*(M)$ is isomorphic to $H_{\mathrm{dR}}^*(N)$. Therefore, if M is a deformation retract of N then $H_{\mathrm{dR}}^*(M) = H_{\mathrm{dR}}^*(N)$. Furthermore, if M is contractible, then $H_{\mathrm{dR}}^*(M) = H_{\mathrm{dR}}^*(\mathrm{pt})$.*

Integration theory and de Rham cohomology are closely related.

Theorem 5.3 *Let M be a compact oriented manifold of dimension m without boundary.*

1. *If ω is an m-form on M with $\int_M \omega \neq 0$, then $0 \neq [\omega] \in H_{\mathrm{dR}}^m(M)$.*
2. $\dim\{H_{\mathrm{dR}}^m(M)\} > 0$.
3. *If ω is a non-vanishing m-form on M, then $0 \neq [\omega] \in H_{\mathrm{dR}}^m(M)$.*

Proof. Let ω be a smooth m-form with $\int_M \omega \neq 0$. If $\omega = d\tilde{\omega}$, then Stokes' Theorem implies $\int_M \omega = \int_{\partial M} \tilde{\omega} = 0$ which is false as, by assumption, $\partial M = \emptyset$. Consequently, $[\omega] \neq 0$ in $H_{\mathrm{dR}}^m(M)$ which proves Assertion 1. Put a Riemannian metric on M and let dvol be the oriented volume form. This is a non-vanishing m-form on M with $\int_M \mathrm{dvol} = \mathrm{Vol}(M)$. Assertion 2 now follows from Assertion 1. If ω is an m-form, we can express $\omega = f \cdot \text{dvol}$. By Theorem 5.1, we may suppose that M is connected; hence since ω is non-vanishing, either f is always positive or f is always negative. Since $\int_M \omega = \int_M f|\text{dvol}|$, $\int_M \omega \neq 0$ and hence $0 \neq [\omega] \in H^m_{\text{dR}}(M)$ by Assertion 1. □

We can now compute the de Rham cohomology groups of the unit sphere S^m in $\mathbb{R}^{m+1}$. Let $\vec{x} = (x^1, \dots, x^{m+1})$ be the usual coordinates on $\mathbb{R}^{m+1}$. Define ω_m on the punctured Euclidean space $\mathbb{R}^{m+1} - \{0\}$ by setting:

$$\omega_m = \|\vec{x}\|^{-m-1} \sum_{i=1}^{m+1} (-1)^{i+1} x^i dx^1 \wedge \dots \wedge \widehat{dx^i} \wedge \dots \wedge dx^{m+1} .$$

The notation $\widehat{dx^i}$ indicates that this element is to be deleted from the wedge product.

Theorem 5.4 *If* $m \geq 1$, $H^p_{\text{dR}}(\mathbb{R}^{m+1} - \{0\}) = H^p_{\text{dR}}(S^m) = \left\{ \begin{array}{ll} \mathbb{R} \cdot [1] & \text{if} \quad p = 0 \\ \mathbb{R} \cdot [\omega_m] & \text{if} \quad p = m \\ 0 & \text{otherwise} \end{array} \right\}$.

Proof. Let N be the north pole and let S be the south pole of S^m. Let

$$\mathcal{O}_1 := S^m - \{N\} \quad \text{and} \quad \mathcal{O}_2 = S^m - \{S\} .$$

Since $\mathcal{O}_i$ is the punctured sphere, $\mathcal{O}_i$ is homotopy equivalent to a point and $\mathcal{O}_1 \cap \mathcal{O}_2$ is diffeomorphic to $S^{m-1} \times (0, \pi)$ which is homotopy equivalent to S^{m-1}. Therefore, the Mayer–Vietoris sequence yields:

$$\begin{array}{ccccccc} H^{p-1}_{\text{dR}}(\mathcal{O}_1) \oplus H^{p-1}_{\text{dR}}(\mathcal{O}_2) & \to & H^{p-1}_{\text{dR}}(\mathcal{O}_1 \cap \mathcal{O}_2) & \to & H^p_{\text{dR}}(S^m) & \to & H^p_{\text{dR}}(\mathcal{O}_1) \oplus H^p_{\text{dR}}(\mathcal{O}_2) \\ \wr\| & & \wr\| & & & & \wr\| \\ H^{p-1}_{\text{dR}}(\text{pt}) \oplus H^{p-1}_{\text{dR}}(\text{pt}) & & H^{p-1}_{\text{dR}}(S^{m-1}) & & & & H^p_{\text{dR}}(\text{pt}) \oplus H^p_{\text{dR}}(\text{pt}) . \end{array}$$

Since $H^p_{\text{dR}}(\text{pt}) = 0$ for $p > 0$, we conclude $H^p_{\text{dR}}(S^m) \simeq H^{p-1}_{\text{dR}}(S^{m-1})$ for $p \geq 2$. We now examine the start of the sequence:

$$0 \to \mathbb{R} \to \mathbb{R} \oplus \mathbb{R} \to H^0_{\text{dR}}(S^{m-1}) \xrightarrow{\mathfrak{d}} H^1_{\text{dR}}(S^m) \to 0 .$$

The sequence simplifies to become:

$$0 \to \mathbb{R} \to H^0_{\text{dR}}(S^{m-1}) \xrightarrow{\mathfrak{d}} H^1_{\text{dR}}(S^m) \to 0 .$$

If $m = 1$, then S^0 is the disjoint union of two points so $H^0_{\text{dR}}(S^0) = \mathbb{R} \oplus \mathbb{R}$. Consequently, $H^1_{\text{dR}}(S^1) = \mathbb{R}$. If $m \geq 2$, then $H^0_{\text{dR}}(S^{m-1}) = \mathbb{R}$ and we conclude $H^1_{\text{dR}}(S^m) = 0$. This establishes the additive structure of $H^*_{\text{dR}}(S^m)$. Since the sphere S^m is a deformation retract of $\mathbb{R}^{m+1} - \{0\}$, we may use Theorem 5.2 to see $H^p_{\text{dR}}(S^m) = H^p_{\text{dR}}(\mathbb{R}^{m+1} - \{0\})$. Therefore, we may identify these two groups. We must now show that the generator of $H^m_{\text{dR}}(S^m)$ is $[\omega_m]$. We show that $d\omega_m = 0$ by computing:

$$
\begin{aligned}
d\omega_m &= d\{\|\vec{x}\|^{-m-1}\} \wedge \sum_{i=1}^{m+1}(-1)^{i+1}x^i dx^1 \wedge \cdots \wedge \widehat{dx^i} \wedge \cdots \wedge dx^{m+1} \\
&\quad + \|\vec{x}\|^{-m-1} d \sum_{i=1}^{m+1}(-1)^{i+1}x^i dx^1 \wedge \cdots \wedge \widehat{dx^i} \wedge \cdots \wedge dx^{m+1} \\
&= -(m+1)\|\vec{x}\|^{-m-3} \sum_{k=1}^{m+1} x^k dx^k \wedge \sum_{i=1}^{m+1}(-1)^{i+1}x^i dx^1 \wedge \cdots \wedge \widehat{dx^i} \wedge \cdots \wedge dx^{m+1} \\
&\quad + \|\vec{x}\|^{-m-1} \sum_{i=1}^{m+1}(-1)^{i+1} dx^i \wedge dx^1 \wedge \cdots \wedge \widehat{dx^i} \wedge \cdots \wedge dx^{m+1} \\
&= -(m+1)\|\vec{x}\|^{-m-3}\|\vec{x}\|^2 dx^1 \wedge \cdots \wedge dx^{m+1} \\
&\quad + (m+1)\|\vec{x}\|^{-m-1} dx^1 \wedge \cdots \wedge dx^{m+1} = 0\,.
\end{aligned}
$$

Since $\|\vec{x}\| = 1$ on S^m, we use Stokes' Theorem to compute:

$$
\begin{aligned}
\int_{S^m} \omega_m &= \int_{S^m} \sum_{i=1}^{m+1}(-1)^{i+1}x^i dx^1 \wedge \cdots \wedge \widehat{dx^i} \wedge \cdots \wedge dx^{m+1} \\
&= \int_{D^{m+1}} d\left\{\sum_{i=1}^{m+1}(-1)^{i+1}x^i dx^1 \wedge \cdots \wedge \widehat{dx^i} \wedge \cdots \wedge dx^{m+1}\right\} \\
&= \int_{D^{m+1}} (m+1) dx^1 \wedge \cdots \wedge dx^{m+1} = (m+1)\operatorname{vol}(D^{m+1}) \neq 0\,.
\end{aligned}
$$

By Theorem 5.3, $[\omega_m] \neq 0$ in $H^m(S^m) = H^m(\mathbb{R}^{m+1} - \{0\})$. □

The outward unit normal on S^m is given by setting $\nu := x^i \partial_{x^i}$. Let int be interior multiplication – see Section 5.2 for details. We then have

$$\omega_m|_{S^m} = \operatorname{int}(\nu)(dx^1 \wedge \cdots \wedge dx^{m+1})\,.$$

The construction is invariant under the action of the special orthogonal group $\mathrm{SO}(m+1)$. Consequently, ω_m is an $\mathrm{SO}(m+1)$ invariant m-form on S^m. Consequently, ω_m is a constant multiple of the oriented volume form on S^m. One evaluates at the north pole to see that the multiple is 1 so in fact ω_m is the oriented volume form. If $r : \mathbb{R}^{m+1} - \{0\} \to S^m$ is the radial retraction $\vec{x} \to \|\vec{x}\|^{-1}\vec{x}$, then on $\mathbb{R}^{m+1} - \{0\}$ we have $\omega_m = r^*(\omega_m|_{S^m})$.

Let $a(\vec{x}) := -\vec{x}$ be the antipodal map of the sphere. Let $\mathbb{Z}_2 := \{\mathrm{Id}, a\}$ act smoothly without fixed points on the sphere S^k for $k \geq 1$. The quotient $S^m/\mathbb{Z}_2$ is the *real projective space* $\mathbb{RP}^m$. We study the behavior of de Rham cohomology under finite regular coverings and compute the cohomology of real projective space as an application.

Theorem 5.5

1. *If Γ is a finite group which acts smoothly and without fixed points on a smooth manifold M, we can let $\tilde{M} := M/\Gamma$ and let $\pi : M \to \tilde{M}$ be the associated regular covering projection. Then $\pi^* : H^p_{\text{dR}}(\tilde{M}) \to H^p_{\text{dR}}(M)$ is injective and*

$$\text{range}\{\pi^*\} = \{[\theta] \in H^p_{\text{dR}}(M) : \sigma^*[\theta] = [\theta] \textit{ for all } \sigma \in \Gamma\}.$$

2. $$H^p_{\text{dR}}(\mathbb{RP}^m) = \left\{ \begin{array}{lll} \mathbb{R} & \textit{if} & p = 0 \\ \mathbb{R} & \textit{if} & p = m \textit{ and } m \textit{ is odd} \\ 0 & \textit{otherwise} & \end{array} \right\}.$$

3. *$\mathbb{RP}^m$ is orientable if and only if m is odd.*

Proof. Since $\pi : M \to \tilde{M}$ is a covering projection, it is a local diffeomorphism. Consequently, if $\tilde{\theta} \in C^\infty(\Lambda^p \tilde{M})$, then $\pi^*\tilde{\theta}$ is invariant under the action of the deck group Γ. The reverse implication also follows and, consequently:

$$\pi^*\{C^\infty(\Lambda^p \tilde{M})\} = \{\theta \in C^\infty(\Lambda^p M) : \gamma^*\theta = \theta \text{ for all } \gamma \in \Gamma\}.$$

Let $\tilde{\theta} \in C^\infty(\Lambda^p \tilde{M})$ satisfy $d\tilde{\theta} = 0$. Suppose $[\pi^*\tilde{\theta}] = 0$ in $H^p_{\text{dR}}(M)$. That means there exists Θ so that $\pi^*\tilde{\theta} = d\Theta$. Since $\pi \circ \gamma = \pi$ for any $\gamma \in \Gamma$, $\gamma^*\pi^* = \pi^*$. This shows that

$$\pi^*\tilde{\theta} = \frac{1}{|\Gamma|}\sum_{\gamma\in\Gamma} \gamma^*\pi^*\tilde{\theta} = \frac{1}{|\Gamma|}\sum_{\gamma\in\Gamma} \gamma^* d\Theta = d\Theta_1 \quad \text{for} \quad \Theta_1 := \frac{1}{|\Gamma|}\sum_{\gamma\in\Gamma} \gamma^*\Theta.$$

If $\delta \in \Gamma$, then

$$\delta^*\Theta_1 = \frac{1}{|\Gamma|}\sum_{\gamma\in\Gamma} \delta^*\gamma^*\Theta = \frac{1}{|\Gamma|}\sum_{\gamma\in\Gamma} (\gamma\delta)^*\Theta = \frac{1}{|\Gamma|}\sum_{\gamma\in\Gamma} \gamma^*\Theta = \Theta_1$$

since the elements $\gamma\delta$ as γ ranges through Γ also parametrize the elements of Γ. Consequently, Θ_1 is invariant under the action of the deck group and can be written in the form $\Theta_1 = \pi^*\tilde{\Theta}_1$. This implies $\pi^*(\tilde{\theta} - d\tilde{\Theta}_1) = 0$. Since π^* is a local diffeomorphism, $\tilde{\theta} - d\tilde{\Theta}_1 = 0$ so $[\theta] = 0$ in $H^p_{\text{dR}}(\tilde{M})$. This shows π^* is injective on cohomology; a similar argument averaging over the deck group shows range$\{\pi^*\}$ is the invariant cohomology. This proves Assertion 1.

We specialize to the natural projection $\pi : S^m \to \mathbb{RP}^m$. Let ω_m be as defined in Theorem 5.4 and let a be the antipodal map. We may compute that $a^*\omega_m = (-1)^{m+1}\omega_m$. By Assertion 1, $H^p_{\text{dR}}(S^m) = 0$ for $p \neq 0, m$. Therefore, $H^p_{\text{dR}}(\mathbb{RP}^m) = 0$ in that range. Furthermore, since S^m is arc connected, $\mathbb{RP}^m$ is arc connected so $H^0_{\text{dR}}(\mathbb{RP}^m) = \mathbb{R}$. Finally, $H^m_{\text{dR}}(\mathbb{RP}^m)$ is isomorphic to the invariant cohomology elements of $H^m(S^m)$. Since $a^*\omega_m = (-1)^{m+1}\omega_m$ and $[\omega_m]$ generates $H^m_{\text{dR}}(S^m)$, $H^m_{\text{dR}}(\mathbb{RP}^m)$ vanishes if m is even and is $\mathbb{R}$ if m is odd. If m is odd, ω_m is invariant under the antipodal map and descends to define an orientation of $\mathbb{R}^m$. If m is even, we apply Theorem 5.3 to see $\mathbb{RP}^m$ is not orientable since $H^m_{\mathrm{dR}}(\mathbb{RP}^m) = \{0\}$. Assertion 3 also follows from Lemma 2.18 in Book I. □

We now determine the additive structure of the de Rham cohomology of complex projective space. We shall determine the ring structure in Theorem 5.18 but postpone that discussion until after we have introduced the first Chern class.

Theorem 5.6 *If* $k \geq 1$, *then* $H^p_{\mathrm{dR}}(\mathbb{CP}^k) = \left\{ \begin{array}{ll} \mathbb{R} & \text{if } 0 \leq p \leq 2k \text{ and } p \text{ is even} \\ 0 & \text{otherwise} \end{array} \right\}.$

Proof. Since $\mathbb{CP}^1 = S^2$, Theorem 5.4 establishes the desired result if $k = 1$. Therefore, we proceed by induction to establish the result in general. We have $\mathbb{CP}^k := S^{2k+1}/S^1$. Let

$$\vec{z} = (z^1, \vec{w}) \in S^{2k+1} \quad \text{for} \quad z^1 \in \mathbb{C} \quad \text{and} \quad \vec{w} \in \mathbb{C}^k .$$

We define:

$$\mathcal{U}_1 := \{(z^1, \vec{w}) \in S^{2k+1} : |z^1| < 2|\vec{w}|\} \quad \text{and} \quad \mathcal{U}_2 := \{(z^1, \vec{w}) \in S^{2k+1} : |z^1| > \tfrac{1}{2}|\vec{w}|\} .$$

These are open subsets of S^{2k+1} which are invariant under the action of S^1; the two open sets $\mathcal{O}_i := \mathcal{U}_i/S^1$ form an open cover of $\mathbb{CP}^k$ to which we will apply Mayer–Vietoris. We set

$$S = (1, 0, \dots, 0) \cdot S^1 \subset S^{2k-1} .$$

Let $f_2(z^1, \vec{w}) := (z^1|z^1|^{-1}, 0) : \mathcal{U}_2 \to S$ and let i_2 be the inclusion of S in $\mathcal{U}_2$. It is then immediate that $f_2 i_2 = \mathrm{Id}_S$. Define a homotopy F_2 from $i_2 f_2$ to $\mathrm{Id}_{\mathcal{U}_2}$ by:

$$F_2(z^1, \vec{w}; t) = (z^1, t\vec{w})(|(z^1, t\vec{w})|)^{-1} .$$

Consequently, $\mathcal{U}_2$ and S are homotopy equivalent spaces. Since the maps in question are S^1 equivariant, $\mathcal{O}_2$ is homotopy equivalent to a point. Next, let

$$\begin{aligned} S^{2k-1} &:= \{(z^1, \vec{w}) \in S^{2k+1} : z^1 = 0\}, \\ f_1(z^1, \vec{w}) &:= (0, \vec{w}/|\vec{w}|) : \mathcal{U}_1 \to S^{2k-1} . \end{aligned}$$

Let i_1 be the inclusion of S^{2k-1} into $\mathcal{U}_1$. The same argument as that given above shows that these equivariant maps descend to define homotopy equivalences between $\mathcal{O}_1$ and

$$\mathbb{CP}^{k-1} = S^{2k-1}/S^1 .$$

Finally, a similar argument shows that $\mathcal{U}_1 \cap \mathcal{U}_2$ is homotopy equivalent to the space

$$\{(z^1, \vec{w}) : |z^1| = |\vec{w}| = \tfrac{1}{\sqrt{2}}\} = S^1 \times S^{2k-1}$$

with the diagonal action $\lambda(\mu, \vec{w}) = (\lambda\mu, \lambda\vec{w})$. Let $T(\mu, \vec{w}) = (\mu, \mu^{-1}\vec{w})$ define a diffeomorphism of $S^1 \times S^{2k-1}$. Since $T(\lambda\mu, \lambda\vec{w}) = (\lambda\mu, \vec{w})$, T intertwines the diagonal action of S^1 on $S^1 \times S^{2k-1}$ with the standard action on S^1 and the trivial action on S^{2k-1}. Dividing by the action of S^1 shows that $\mathcal{O}_1 \cap \mathcal{O}_2$ is homotopy equivalent to S^{2k-1}. Therefore, we have

$$H^p_{\mathrm{dR}}(\mathcal{O}_1) = H^p_{\mathrm{dR}}(\mathbb{CP}^{k-1}), \quad H^p_{\mathrm{dR}}(\mathcal{O}_2) = H^p_{\mathrm{dR}}(\mathrm{pt}), \quad H^p_{\mathrm{dR}}(\mathcal{O}_1 \cap \mathcal{O}_2) = H^p_{\mathrm{dR}}(S^{2k-1}).$$

Consequently, the Mayer–Vietoris sequence becomes:

$$\begin{array}{ccccccc} \cdots H^{p-1}_{\mathrm{dR}}(\mathcal{O}_1 \cap \mathcal{O}_2) & \to & H^p_{\mathrm{dR}}(\mathbb{CP}^k) & \to & H^p_{\mathrm{dR}}(\mathcal{O}_1) \oplus H^p_{\mathrm{dR}}(\mathcal{O}_2) & \to & H^p_{\mathrm{dR}}(\mathcal{O}_1 \cap \mathcal{O}_2) \cdots \\ \wr\| & & & & \wr\| & & \wr\| \\ H^{p-1}_{\mathrm{dR}}(S^{2k-1}) & & & & H^p_{\mathrm{dR}}(\mathbb{CP}^{k-1}) \oplus H^p_{\mathrm{dR}}(\mathrm{pt}) & & H^p_{\mathrm{dR}}(S^{2k-1}). \end{array}$$

We examine the beginning of the sequence and take $p = 0$. The spaces in question are path connected so $H^0_{\mathrm{dR}}(\cdot) = \mathbb{R}$. Furthermore, by induction $H^1_{\mathrm{dR}}(\mathbb{CP}^{k-1}) = 0$ so the sequence is:

$$0 \to \mathbb{R} \to \mathbb{R} \oplus \mathbb{R} \to \mathbb{R} \to H^1_{\mathrm{dR}}(\mathbb{CP}^k) \to 0.$$

It now follows that $H^1_{\mathrm{dR}}(\mathbb{CP}^k) = \{0\}$. If $1 \le q < 2k - 1$, then $H^q_{\mathrm{dR}}(\mathrm{pt}) = H^q_{\mathrm{dR}}(S^{2k-1}) = 0$ and the Mayer–Vietoris sequence becomes: $0 \to H^p_{\mathrm{dR}}(\mathbb{CP}^k) \to H^p_{\mathrm{dR}}(\mathbb{CP}^{k-1}) \to 0$ and, by induction,

$$H^p_{\mathrm{dR}}(\mathbb{CP}^k) = \left\{ \begin{array}{lll} 0 & \text{if} \quad 1 \le p < 2k-1 & \text{is odd} \\ \mathbb{R} & \text{if} \quad 1 \le p < 2k-1 & \text{is even} \end{array} \right\}.$$

Finally, we examine the top of the sequence using the fact that $H^p_{\mathrm{dR}}(\mathbb{CP}^{k-1}) = 0$ for $p = 2k$ and $p = 2k - 1$:

$$0 \to H^{2k-1}_{\mathrm{dR}}(\mathbb{CP}^k) \to 0 \to H^{2k-1}_{\mathrm{dR}}(S^{2k-1}) \to H^{2k}_{\mathrm{dR}}(\mathbb{CP}^k) \to 0.$$

It now follows $H^{2k}_{\mathrm{dR}}(\mathbb{CP}^k) = \mathbb{R}$ and $H^{2k-1}_{\mathrm{dR}}(\mathbb{CP}^k) = 0$. □

5.1.2 SIMPLE COVER. We say that a finite open cover $\mathcal{U} = \{\mathcal{O}_i\}_{i \in A}$ of a topological space X is a *simple cover* if for any subset B of the indexing set A, the intersection $\cap_{i \in B} \mathcal{O}_i$ is either contractible or empty.

Theorem 5.7 *Let M be a manifold of dimension m.*

1. *If M is compact, then there exists a finite simple cover of M.*
2. *If M admits a finite simple cover, then $H^q_{\mathrm{dR}}(M)$ is finite-dimensional for any q.*

Proof. Put an auxiliary Riemannian metric g on M. By Lemma 3.14 in Book I, small geodesic balls are geodesically convex. If M is compact, we may cover M by a finite collection of small geodesic balls. Since the intersection of geodesically convex sets is geodesically convex (and hence contractible) or empty, this shows that M admits finite simple covers.

Let $\{\mathcal{O}_i\}_{1\le i\le \ell}$ be a finite simple cover of M. We proceed by induction on ℓ to show $H^*_{\mathrm{dR}}(M)$ is finite-dimensional. If $\ell = 1$, the result follows since $M = \mathcal{O}_1$ is contractible and hence homotopy equivalent to a point. If $\ell > 1$, we let

$$M_1 := \cup_{i<\ell}\mathcal{O}_i, \quad M_2 = \mathcal{O}_\ell, \quad M_3 := M_1 \cap M_2 = \cup_{i<\ell}\{\mathcal{O}_i \cap \mathcal{O}_\ell\}.$$

Since M_1, M_2, and M_3 admit simple covers with at most $\ell - 1$ elements, we can apply induction to see $H^*_{\mathrm{dR}}(M_i)$ is finite for $i = 1, 2, 3$. The desired result now follows from the Mayer–Vietoris sequence in de Rham cohomology. □

5.1.3 EXAMPLES.

1. Let $M_1 := \mathbb{R}^2 - \{(0,0)\}$ be the plane with one puncture. Let

$$\mathcal{O}_1 := \{(x,y) \in M_1 : x > 0\}, \qquad \mathcal{O}_2 := \{(x,y) \in M_1 : x < 0\},$$
$$\mathcal{O}_3 := \{(x,y) \in M_1 : y > 0\}, \qquad \mathcal{O}_4 := \{(x,y) \in M_1 : y < 0\}.$$

   Then $\mathcal{O}_i \cap \mathcal{O}_j$ is empty for $(i,j) = (1,2)$ or $(i,j) = (3,4)$ whilst otherwise the intersection is one of the four open quadrants in the plane. Consequently, this is a simple cover of M_1 and M_1 is a non-compact manifold which admits a finite simple cover. Note that the circle $x^2 + y^2 = 1$ is a deformation retract of M_1 so $H^*_{\mathrm{dR}}(M_1) = H^*_{\mathrm{dR}}(S^1)$.

2. Let $M_2 := \mathbb{R}^2 - \cup_{n\in\mathbb{N}}\{(n,0)\}$ be the plane with a countable number of punctures located at discrete points. Let

$$\omega_n := \frac{-y\,dx + (x-n)\,dy}{(x-n)^2 + y^2}\,.$$

   One verifies that $d\omega_n = 0$ and that for $\epsilon < 1$

$$\int_{(x-n)^2+y^2=\epsilon^2} \omega_k = \left\{ \begin{array}{ll} 0 & \text{if } k \neq n \\ 2\pi & \text{if } k = n \end{array} \right\}.$$

   We may then use Stokes' Theorem to see that $\{[\omega_n]\}_{n\in\mathbb{N}}$ are linearly independent elements of $H^1_{\mathrm{dR}}(M_2)$ and hence $H^1_{\mathrm{dR}}(M_2)$ is not finite-dimensional. Consequently, M_2 does not admit a finite simple cover.

5.2 CLIFFORD ALGEBRAS

The following algebraic structure was first examined by the English mathematician William Kingdon Clifford and by the German mathematician Rudolf Otto Sigismund Lipschitz.

W. Clifford (1845–1879)

R. Lipschitz (1832–1903)

Throughout Section 5.2, let $(V, (\cdot,\cdot))$ be a positive definite inner product space of dimension m. If $\{e^1,\ldots,e^m\}$ is an orthonormal basis for V, and if $I = (i_1,\ldots,i_p)$ is a collection of strictly increasing indices $1 \le i_1 < \cdots < i_p \le r$, we define $e^I := e^{i_1} \wedge \cdots \wedge e^{i_p} \in \Lambda^p(V)$. Extend the inner product to the entire tensor algebra and in particular to the space of p-forms $\Lambda^p(V)$. The $\{e^I\}_{|I|=p}$ then form an orthonormal basis for $\Lambda^p(V)$. If $v \in V$, let

$$\mathrm{ext}(v) : \omega \to v \wedge \omega \quad \text{for} \quad \omega \in \Lambda(V)$$

be *exterior multiplication*. Let $\mathrm{int}(v)$ be the dual, *interior multiplication*; this is characterized by the identity:

$$(\mathrm{ext}(v)\omega, \phi) = (\omega, \mathrm{int}(v)\phi)\,.$$

Let $c(v) := \mathrm{ext}(v) - \mathrm{int}(v)$ be *Clifford multiplication.*

Lemma 5.8 Adopt the notation established above. Then $c(v)^2 = -\|v\|^2\,\mathrm{Id}$ and

$$\begin{aligned} &c(v)c(w) + c(w)c(v) = -2(v,w)\,\mathrm{Id}, && \mathrm{ext}(v)\,\mathrm{int}(w) + \mathrm{int}(w)\,\mathrm{ext}(w) = -2(v,w)\,\mathrm{Id},\\ &\mathrm{ext}(v)\,\mathrm{ext}(w) + \mathrm{ext}(w)\,\mathrm{ext}(v) = 0, && \mathrm{int}(v)\,\mathrm{int}(w) + \mathrm{int}(w)\,\mathrm{int}(w) = 0\,. \end{aligned}$$

Proof. Let $\{e^1,\ldots,e^m\}$ be an orthonormal basis for V. Then

$$\mathrm{ext}(e^1)e^I = \left\{\begin{array}{ll} e^J \text{ for } J = I \cup \{1\} & \text{if } i_1 > 1 \\ 0 & \text{if } i_1 = 1 \end{array}\right\},$$

$$\mathrm{int}(e^1)e^I = \left\{\begin{array}{ll} 0 & \text{if } i_1 > 1 \\ e^J \text{ for } J = I - \{1\} & \text{if } i_1 = 1 \end{array}\right\}.$$

This shows that exterior multiplication by e^1 adds the index 1 if possible while interior multiplication by e^1 cancels the index 1 if possible. It is now immediate that $c(e_1)^2 = -\,\mathrm{Id}$. Given an arbitrary vector, we can always choose an orthonormal frame so $v = \|v\|e^1$ and hence, by rescaling,

$$c(v)^2 = -\|v\|^2\,\mathrm{Id}\,.$$

Polarizing this identity and then decomposing Clifford multiplication into its component graded pieces yields the remaining identities. □

Fix an orientation of V and let dvol be the *oriented volume form*. If $\{e^1, \ldots, e^m\}$ is an oriented orthonormal basis for V, then $\mathrm{dvol} = e^1 \wedge \cdots \wedge e^m$. If $\{f^1, \ldots, f^m\}$ is an arbitrary oriented basis for V, let $\epsilon^{ij} := (f^i, f^j)$. We then have that

$$\mathrm{dvol} = \det(\epsilon)^{-1/2} f^1 \wedge \cdots \wedge f^m .$$

The *Hodge* $\star$ *operator* is the linear map from $\Lambda^p(V)$ to $\Lambda^{m-p}(V)$ characterized by the relation:

$$\omega_p \wedge \star_p \phi_p = g(\omega_p, \phi_p)\,\mathrm{dvol} \quad \text{for all} \quad \omega_p, \phi_p \in \Lambda^p(V) .$$

For example, $\star(e^1 \wedge \cdots \wedge e^p) = e^{p+1} \wedge \cdots \wedge e^m$. This shows

$$\star_{m-p} \star_p = (-1)^{p(m-p)} .$$

The *Clifford algebra*, $\mathrm{Clif}(V, (\cdot,\cdot))$, is the universal unital algebra generated by V subject to the *Clifford commutation relations*

$$v \star w + w \star v = -2(v, w)\,\mathrm{Id} .$$

We use Lemma 5.8 to see that Clifford multiplication gives a well-defined algebra morphism $c : \mathrm{Clif}(V, (\cdot,\cdot)) \to \mathrm{Hom}(V)$. The map

$$\Xi_c : \omega \to c(\omega) \cdot 1 \tag{5.2.a}$$

defines a linear map from $\mathrm{Clif}(V, (\cdot,\cdot))$ to $\Lambda(V)$. The Clifford commutation relations show that the elements $\{e^{i_1} \star \cdots \star e^{i_p}\}$ for $1 \le i_1 < \cdots < i_p \le m$ and $0 \le p \le m$ are a linear spanning set for $\mathrm{Clif}(V, (\cdot,\cdot))$ (if $p = 0$, the empty product is 1). We show that Ξ_c is a natural (i.e., basis independent) linear isomorphism from $\mathrm{Clif}(V, (\cdot,\cdot))$ to $\Lambda(V)$ by computing

$$c(e^{i_1} \star \cdots \star e^{i_p})1 = e^{i_1} \wedge \cdots \wedge e^{i_p} \quad \text{for} \quad 1 \le i_1 < \cdots < i_p \le m .$$

We emphasize that the algebra structure is not preserved by Ξ_c since, for example, one has $\Xi_c(e^1 \star e^1) = \Xi_c(-1) = -1$ while $\Xi_c(e^1) \wedge \Xi_c(e^1) = e^1 \wedge e^1 = 0$. By an abuse of notation, we shall also let the oriented volume form, dvol, be the associated element $\mathrm{Clif}(V, (\cdot,\cdot))$; $c(\mathrm{dvol})$ is then a natural endomorphism of the exterior algebra $\Lambda(V)$.

Lemma 5.9 Let $(V, (\cdot,\cdot))$ be an oriented positive definite inner product space of dimension m. Then $c(\mathrm{dvol})^2 = (-1)^{m(m+1)/2}\,\mathrm{Id}$ and $\star_p = (-1)^{(2m-p+1)p/2} c(\mathrm{dvol})$ on $\Lambda^p(V)$.

Proof. Let $\{e^1, \ldots, e^m\}$ be an oriented orthonormal basis for V. We establish the first identity by computing:

$$\begin{aligned}
\mathrm{dvol}^2 &= e^1 \star \cdots \star e^m \star e^1 \star \cdots \star e^m = (-1)^m e^2 \star \cdots \star e^m \star e^2 \star \cdots \star e^m \\
&= \cdots = (-1)^{m+(m-1)+\cdots+1}\,\mathrm{Id} = (-1)^{m(m+1)/2}\,\mathrm{Id}, \\
c(\mathrm{dvol})^2 &= c(\mathrm{dvol}^2) = (-1)^{m(m+1)/2}\,\mathrm{Id} .
\end{aligned}$$

We compare $\star\omega_p$ and $c(\mathrm{dvol})\omega_p$. We let $\omega_p = e^I$ where $|I| = p$. By permuting the elements in the basis, it is sufficient to consider the special case in which $I = \{1, \dots, p\}$. We use the isomorphism Ξ_c of Equation (5.2.a) to replace $\Lambda(V)$ by $\mathrm{Clif}(V, (\cdot,\cdot))$ and complete the proof by computing:

$$\begin{aligned}
&\Xi_c^{-1}\{\star(e^1 \wedge \dots \wedge e^p)\} = \Xi_c^{-1}\{e^{p+1} \wedge \dots \wedge e^m\} = e^{p+1} \star \dots \star e^m, \\
&\Xi_c^{-1}\{c(\mathrm{dvol})e^1 \wedge \dots \wedge e^p\} \\
&\quad = e^1 \star \dots \star e^m \star e^1 \star \dots \star e^p = (-1)^{m+(m-1)+\dots+(m-p+1)} e^{p+1} \star \dots \star e^m \\
&\quad = (-1)^{(2m-p+1)p/2} \Xi_c^{-1}(e^1 \wedge \dots \wedge e^p). \qquad \square
\end{aligned}$$

The following is a technical result that will play an important role in our discussion of the Maurer–Cartan forms subsequently in Theorem 6.30. Let $\mathrm{U}(k)$ be the k-dimensional unitary group.

Lemma 5.10 Let m be even. There exists a map $g : S^{m-1} \to \mathrm{U}(k)$ so the $m-1$ form $\mathrm{Tr}\{(g^{-1}dg)^{m-1}\}$ is a non-zero constant multiple of $\mathrm{dvol}_{S^{m-1}}$.

Proof. If $m = 2$, we could define $g(x^1, x^2) := x^1 + \sqrt{-1}x^2$ and if $m = 4$, we could define

$$g(x^1, x^2, x^3, x^4) := \begin{pmatrix} x^1 + \sqrt{-1}x^2 & x^3 + \sqrt{-1}x^4 \\ -x^3 + \sqrt{-1}x^4 & x^1 - \sqrt{-1}x^2 \end{pmatrix}$$

and verify the conclusion of the Lemma directly. For general m, however, it is convenient to use Clifford algebras. Our construction is closely related to Bott periodicity and we refer to Atiyah, Bott, and Shapiro [3] for further details; we mention it only to warn the reader that we are entering into deep waters.

Let $(V, \langle\cdot,\cdot\rangle)$ be an m-dimensional positive definite inner product space. We could take $V = \mathbb{R}^m$ with the usual inner product, but it is convenient to work in a coordinate free setting for the moment. Let c be Clifford multiplication. Then dc is a 1-form valued endomorphism of $\Lambda(V)$. Fix an orientation of V and let dvol be the oriented volume form. Define a 1-form valued endomorphism of $\Lambda(V)$ by setting $\theta(v) := c(v)dc(v)$. Let v_0 be a unit vector in V. Choose an oriented orthonormal frame $\{e_i\}$ for V so that $v_0 = e_1$. Let $x = x^i e_i$ define the dual coordinates on V. We have

$$\begin{aligned}
&x(v_0) = (1, 0, \dots, 0), \quad dx^1(v_0)|_{S^{m-1}} = 0, \quad \mathrm{dvol}_V(v_0) = dx^1 \wedge \dots \wedge dx^m, \\
&\mathrm{dvol}_{S^{m-1}}(v_0) = dx^2 \wedge \dots \wedge dx^m, \quad dc(v_0)|_{S^{m-1}} = c(e_2)dx^2 + \dots + c(e_m)dx^m, \\
&c(v_0) = e_1, \quad \text{and} \quad \theta(v_0)|_{S^{m-1}} = c(e_1)c(e_2)dx^2 + \dots + c(e_1)c(e_m)dx^m.
\end{aligned}$$

Let $2 \le i, j \le m$. The commutation relations

$$\begin{aligned} c(e_1)c(e_i)c(e_1)c(e_j) &= \left\{\begin{array}{ll} -c(e_1)c(e_j)c(e_1)c(e_i) & \text{if } i \ne j \\ +c(e_1)c(e_j)c(e_1)c(e_i) & \text{if } i = j \end{array}\right\} \\ dx^i \wedge dx^j &= \left\{\begin{array}{ll} -dx^j \wedge dx^i & \text{if } i \ne j \\ 0 & \text{if } i = j \end{array}\right\} \end{aligned}$$

imply $\theta(v_0)^{m-1}|_{S^{m-1}} = (m-1)!c(e_1)c(e_2)\cdot \dots \cdot c(e_1)c(e_{m-1}) \otimes \text{dvol}_{S^{m-1}}$. If $2 \le i < m$, then

$$c(e_1)c(e_i)c(e_1)c(e_{i+1}) = -c(e_1)c(e_1)c(e_i)c(e_{i+1}) = c(e_i)c(e_{i+1}).$$

Since $m-1$ is odd, we may use this relation to show

$$\begin{aligned} \theta(v_0)^{m-1}|_{S^{m-1}} &= (m-1)!c(e_1)c(e_2)\cdot \dots \cdot c(e_m) \otimes \text{dvol}_{S^{m-1}}(v_0) \\ &= (m-1)!c(\text{dvol}_V(v_0)) \otimes \text{dvol}_{S^{m-1}}(v_0) \end{aligned}$$

is independent of the particular unit vector v_0 which was chosen. The parity of m mod 4 now enters as

$$c(\text{dvol})^2 = \left\{\begin{array}{ll} 1 & \text{if } m \equiv 0 \mod 4 \\ -1 & \text{if } m \equiv 2 \mod 4 \end{array}\right\}.$$

If $m \equiv 0 \mod 4$, let $\epsilon = +1$; if $m \equiv 2 \mod 4$, let $\epsilon = \sqrt{-1}$. Consequently, we may decompose $\Lambda(V) \otimes_{\mathbb{R}} \mathbb{C} = \Lambda^+ \oplus \Lambda^-$ into the eigenspaces of $c(\text{dvol})$ where

$$\Lambda^\pm := \{\omega \in \Lambda(V) \otimes_{\mathbb{R}} \mathbb{C} : c(\text{dvol})\omega = \pm\epsilon\omega\}.$$

Let $v \in V$. Since m is even, $c(v)$ anti-commutes with $c(\text{dvol})$ and interchanges Λ^+ with Λ^-. Therefore, we may decompose

$$c(v) = c_+(v) + c_-(v) \quad \text{where} \quad c_\pm(v) : \Lambda^\pm \to \Lambda^\mp, \quad \text{and} \quad \theta(v) = \theta_+(v) + \theta_-(v).$$

Note that $\theta_\pm(v) := c_\mp(v)dc_\pm(v)$ belongs to the vector space $\text{Hom}(\Lambda^\pm, \Lambda^\pm) \otimes \Lambda^1(V)$. The computation performed above then yields

$$\text{Tr}\{\theta_\pm(v)^{m-1} \text{ on } \Lambda^\pm(V)\} = \pm(m-1)!\epsilon \dim\{V^\pm\}\,\text{dvol} \quad \text{if} \quad v \in S^{m-1}.$$

Clifford multiplication preserves the inner product and hence is unitary. Let T be a fixed unitary map from Λ^- to Λ^+. Let $v \in V$ be an arbitrary unit vector, and let $g(v) = Tc(v)$ define a smooth map from S^{m-1} to the unitary group $\text{U}(\Lambda^+)$. Then

$$g^{-1}dg = c_+(v)^{-1}T^{-1}Tc_+(v) = c_+(v)^{-1}dc_+(v) = -c_-(v)dc_+(v) = -\theta_+(v)$$

so $\text{Tr}\{(g^{-1}dg)^{m-1}\}$ is a non-zero constant multiple of dvol as desired. □

5.3 THE HODGE DECOMPOSITION THEOREM

The material of this section was introduced by the Italian mathematician E. Beltrami, the Scottish mathematician W. Hodge, and the French mathematician P. Laplace.

E. Beltrami (1835–1899) W. Hodge (1903–1975) P. Laplace (1749–1827)

Let (M, g) be a compact Riemannian manifold. We extend g to a positive definite inner product on tensors of all types. We integrate this inner product with respect to the Riemannian measure $|\,\text{dvol}\,|$ to define a global positive definite inner product on $C^\infty(\Lambda^p M)$ and let $L^2(\Lambda^p M)$ be the L^2 completion. Let

$$d : C^\infty(\Lambda^p M) \to C^\infty(\Lambda^{p+1}) \quad \text{and} \quad \delta : C^\infty(\Lambda^{p+1}) \to C^\infty(\Lambda^p M)$$

be exterior differentiation and the L^2 adjoint, respectively. The Laplacian is then given by

$$\Delta := d\delta + \delta d\,.$$

This is also often called the *Hodge–Beltrami Laplacian*. We will express these operators in terms of the Levi–Civita connection in Theorem 5.14 presently; we suppress indices to simplify the notation.

The Laplacian is a self-adjoint operator. The spectral theory of a self-adjoint operator in finite dimensions is very simple; it is diagonalizable. The Laplacian is an operator on an infinite-dimensional space and the corresponding theory is more complicated and beyond the scope of this book. We refer instead to Gilkey [20] noting that there are many excellent treatments of this subject. If $\phi \in C^\infty(\Lambda^p M)$, define the C^k-norm by setting:

$$\|\phi\|_{C^k} := \sup_{x \in M} \sum_{\ell \le k} \|\nabla^\ell \phi\|(x)\,.$$

Theorem 5.11 *Let Δ^p be the Laplacian acting on p-forms on a compact Riemannian manifold (M, g) of dimension m.*

1. *There exists a complete orthonormal basis $\{\phi_n\}$ for $L^2(\Lambda^p M)$ where $\phi_n \in C^\infty(\Lambda^p M)$ satisfies $\Delta^p \phi_n = \lambda_n \phi_n$. The eigenvalues (repeated according to multiplicity) form a discrete set tending to infinity and may be ordered so $0 \le \lambda_1 \le \lambda_2 \cdots$. Given $\epsilon > 0$ and $k \ge 0$, there exists an integer $\ell = \ell(k, m)$ and $N = N(\epsilon, p, M, k)$ so that if $n \ge N$, then:*

$$n^{\frac{2}{m} - \epsilon} < \lambda_n < n^{\frac{2}{m} + \epsilon} \quad \text{and} \quad \|\phi_n\|_{C^k} \le n^\ell\,.$$

2. *Let $\gamma_n(\phi) := (\phi, \phi_n)_{L^2}$ be the Fourier coefficients for $\phi \in L^2(\Lambda^p M)$. Then the coefficients are "rapidly decreasing" ($\lim_{n\to\infty} n^k \gamma_n(\phi) = 0$ for every k) if and only if $\phi \in C^\infty(\Lambda^p M)$. In this setting, the expansion $\phi = \sum_n \gamma_n(\phi)\phi_n$ of ϕ as a generalized Fourier series converges uniformly in the C^k topology for all k.*

In the case of the circle, the decomposition of Assertion 2 is the usual decomposition in terms of Fourier series which was first introduced by the French mathematician J. Fourier [19] and later made more precise by the German mathematicians P. Dirichlet and G. Riemann.

P. Dirichlet (1805–1859) J. Fourier (1768–1830) G. Riemann (1826–1866)

Let $E(\lambda, \Delta^p) := \{\phi \in C^\infty(\Lambda^p M) : \Delta^p \phi = \lambda\phi\}$ be the associated eigenspaces. If $\operatorname{Spec}(\Delta^p)$ is the collection of eigenvalues, then we have a complete orthogonal decomposition

$$L^2(\Lambda^p M) = \oplus_{\lambda \in \operatorname{Spec}(\Delta^p)} E(\lambda, \Delta^p).$$

The eigenspaces $E(\lambda, \Delta^p)$ are finite-dimensional representation spaces for the group of isometries of M; this will play an important role in our proof of the Peter–Weyl Theorem (Theorem 6.16) subsequently.

5.3.1 SPHERICAL HARMONICS. We do not want to go too deeply into this subject. Nevertheless, it is worth discussing the spectral resolution of the Laplacian on the sphere to illustrate the phenomena involved. Let $\vec{x} = (x^0, \dots, x^m) \in \mathbb{R}^{m+1}$, and let S^m be the unit sphere. Let $\Delta_e := -\partial^2_{x^0} - \cdots - \partial^2_{x^m}$ be the Euclidean Laplacian, and let Δ_s be the spherical Laplacian. Let $S(m+1, j)$ be the vector space of polynomials which are homogeneous of degree j in $m+1$ variables and let $H(m+1, j)$ be the subspace of harmonic polynomials which are homogeneous of degree j in $m+1$ variables:

$$S(m+1, j) := \{f \in \mathbb{R}[x^0, \dots, x^m] : f(t\vec{x}) = t^j f(\vec{x}) \text{ for } t \in \mathbb{R}\},$$
$$H(m+1, j) := \{f \in S(m+1, j) : \Delta_e f = 0\}.$$

For example, if $r^2 := \|\vec{x}\|^2$, then $r^2 \in S(2, j)$, $x^i \in H(m+1, 1)$, and $x^0 x^1 \in H(m+1, 2)$.

Theorem 5.12 *Let $m \geq 1$.*

1. *If $j \in \mathbb{N}$, then* $\dim\{S(m+1, j)\} = \binom{m+j}{m}$.

2. *If* $j \in \mathbb{N}$, *then* $S(m+1, j) = r^2 S(m+1, j-2) \oplus H(m+1, j)$.
3. *If* $j \in \mathbb{N}$, *then* $\dim\{H(m+1, j)\} = \binom{m+j}{m} - \binom{m+j-2}{m}$.
4. *We have that* $\{j(j+m-1), H(m+1, j)\}_{j\in\mathbb{N}}$ *is the discrete spectral resolution of the Laplacian* Δ_{S^m} *on* S^m, *i.e.,* $\lambda_j := j(j+m-1)$ *for* $j = 0, 1, 2, \ldots$ *is an eigenvalue of* Δ_{S^m} *with associated eigenspace being the restriction of the functions in* $H(m+1, j)$ *to the sphere, and there are no other eigenvalues.*

Proof. It is clear that $S(m+1, j) = x^{m+1} \cdot S(m+1, j-1) \oplus S(m, j)$. Assertion 1 follows from the resulting recursion relations:

$$\begin{aligned}&\dim\{S(m+1, j)\} = \dim\{S(m+1, j-1)\} + \dim\{S(m, j)\},\\ &\dim\{S(m+1, 0)\} = 1 \quad \text{and} \quad \dim\{S(1, j)\} = 1\,.\end{aligned}$$

We decompose a homogeneous polynomial $p \in S(m+1, j)$ in the form

$$p = \sum_\alpha p_\alpha x^\alpha \quad \text{for} \quad \alpha = (a_1, \ldots, a_{m+1}) \quad \text{and} \quad x^\alpha := x^{a_1} \cdot \ldots \cdot x^{a_{m+1}}\,.$$

Let $P(p) := \sum_\alpha p_\alpha \partial_\alpha$ be the associated partial differential operator. Define a Euclidean inner product $\langle \cdot, \cdot \rangle$ on $S(m+1, j)$ by setting:

$$\langle p, q \rangle := P(p)(q) = \sum_{\alpha,\beta} p_\alpha \partial_\alpha \{q_\beta x^\beta\} = \sum_\alpha \alpha! p_\alpha q_\alpha\,.$$

Let $\Delta_{\mathbb{R}^{m+1}} = -\partial^2_{x^1} - \cdots - \partial^2_{x^{m+1}}$ be the Laplacian on $\mathbb{R}^{m+1}$. Since $P(r^2) = -\Delta_{\mathbb{R}^{m+1}}$, we have the crucial intertwining relationship $-\langle p, \Delta_{\mathbb{R}^{m+1}} q\rangle = \langle r^2 p, q\rangle$. Multiplication by r^2 is injective. Since $\operatorname{coker}(r^2) = \ker\{\Delta_{\mathbb{R}^{m+1}}\}$, this proves Assertion 2 and Assertion 3. Identify a homogeneous harmonic function with its restriction to S^m. Let $\mathcal{A}$ be the subspace of $C^\infty(S^m)$ which is generated additively by the spaces $H(m+1, j)$. Since $r^2|_{S^m} = 1$, Assertion 2 implies:

$$\begin{aligned}&\sum_{\nu \le 2j} H(m+1, \nu) = \{S(m+1, 2j) + S(m+1, 2j-1)\}|_{S^m},\\ &\mathcal{A} = \cup_j \{S(m+1, 2j) + S(m+1, 2j-1)\}|_{S^m}\,.\end{aligned}$$

Since $S(m+1, j) \cdot S(m+1, k) \subset S(m+1, j+k)$ and since $1 \in S(m+1, 0)$, $\mathcal{A}$ is a unital subalgebra of $C^\infty(S^m)$. As the coordinate functions $x^i \in H(m+1, 1)$, $\mathcal{A}$ separates points of S^m. Consequently, by the Stone–Weierstrass Theorem, $\mathcal{A}$ is dense in $C^\infty(S^m)$ and hence in $L^2(S^m)$.

Introduce parameters $x = (r, \theta)$ on $\mathbb{R}^{m+1}$ for $r \in (0, \infty)$ and $\theta \in S^m$ and express the Laplacian in the form $\Delta_{\mathbb{R}^{m+1}} = -\partial_r^2 - mr^{-1}\partial_r + r^{-2}\Delta_{S^m}$. If $f \in H(m+1, j)$, then $\Delta_{\mathbb{R}^{m+1}}(f) = 0$ so

$$\Delta_{S^m} f(\theta) = j(j+m-1) f(\theta) \quad \text{for} \quad f \in H(m+1, j)\,.$$

Since Δ_{S^m} is self-adjoint, $E(\lambda, \Delta_{S^m}) \perp E(\mu, \Delta_{S^m})$ for $\lambda \neq \mu$. Since

$$H(m+1, \nu) \subseteq E(j(j+m-1), \Delta_{S^m}),$$

the spaces $H(m+1, j)$ and $H(m+1, k)$ are orthogonal in $L^2(S^m)$ for $j \neq k$. The desired result now follows as

$$L^2(S^m) = \oplus_j H(m+1, j) \subset \oplus_j E(j(j+m-1), \Delta_{S^m}) \subset L^2(S^m). \qquad \square$$

If $m = 1$, Theorem 5.12 gives rise to the usual Fourier series decomposition of $L^2(S^1)$. We complexify. The map $\theta \to \cos\theta + \sqrt{-1}\sin\theta$ introduces coordinates on S^1 where θ is the usual periodic parameter; $\Delta_s = -\partial_\theta^2$. Let $\phi_n(\theta) = e^{\sqrt{-1}n\theta}/\sqrt{2\pi}$. Set $z = x^0 + \sqrt{-1}x^1$. Then

$$\phi_n(\theta) = z^n \quad \text{for} \quad n \geq 0 \quad \text{and} \quad \phi_n(\theta) = \bar{z}^n \quad \text{for} \quad n < 0.$$

Since holomorphic and anti-holomorphic functions are harmonic, $H(2, n) = \text{span}_{\mathbb{C}}\{z^n, \bar{z}^n\}$. Furthermore, the spectral resolution of the scalar Laplacian on the circle is given by $\{(2\pi)^{-1/2}\phi_n, n^2\}_{n\in\mathbb{Z}}$ where the constant function 1 spans the kernel of the Laplacian. The decomposition $\phi = \sum_n \gamma_n(\phi)\phi_n$ of Theorem 5.11 then becomes the usual decomposition of ϕ into Fourier series. More generally, one may consider the m-dimensional torus $\mathbb{T}^m := S^1 \times \cdots \times S^1$. Set

$$\vec{n} = (n_1, \ldots, n_m), \quad \vec{\theta} = (\theta_1, \ldots, \theta_m), \quad \phi_n = e^{\sqrt{-1}\vec{n}\cdot\vec{\theta}}$$

to obtain a complete spectral resolution of the scalar Laplacian $\Delta_{\mathbb{T}^m} = -\partial_{\theta_1}^2 - \cdots - \partial_{\theta_n}^2$ as

$$\{(2\pi)^{-m/2}\phi_{\vec{n}}, \|\vec{n}\|^2\}_{\vec{n}\in\mathbb{Z}^m}.$$

5.3.2 THE HODGE DECOMPOSITION THEOREM [32].

The following result of W. Hodge expresses the de Rham cohomology in terms of the kernel of the Laplacian. We caution the reader that the ring structure is not captured by the harmonic forms; the wedge product of two harmonic forms need not be harmonic. The finiteness of the cohomology groups also follows from Theorem 5.7 since a compact manifold admits a finite simple cover.

Theorem 5.13 *Let M be a compact Riemannian manifold. Fix p. There is an L^2 orthogonal direct sum decomposition:*

$$C^\infty(\Lambda^p M) = \ker\{\Delta^p\} \oplus d\{C^\infty(\Lambda^{p-1}M)\}, \oplus\delta\{C^\infty(\Lambda^{p+1}M)\},$$

where $\ker\{\Delta^p\} = \ker\{d\} \cap \ker\{\delta\}$ *is finite-dimensional. The map* $\omega \to [\omega]$ *defines a natural isomorphism between* $\ker\{\Delta^p\}$ *and* $H^p_{\text{dR}}(M)$.

Proof. We suppress the index p in the interests of notational simplification for the remainder of the proof. We use Theorem 5.11 to see that the eigenvalues of the Laplacian tend to infinity; this implies that the kernel is finite-dimensional. If ϕ is a smooth differential form, then we may expand $\phi = \sum_n \sigma_n(\phi)\phi_n$ in a generalized Fourier series where the convergence is uniform in the C^k topology for any k. Decompose $\phi = \Phi_0 + \Phi_1$ for

$$\Phi_0 = \sum_{n:\lambda_n=0} \sigma_n(\phi)\phi_n \quad\text{and}\quad \Phi_1 = \sum_{n:\lambda_n\neq 0} \sigma_n(\phi)\phi_n = \Delta \sum_{n:\lambda_n\neq 0} \sigma_n(\phi)\lambda_n^{-1}\phi_n\,.$$

Therefore, $\Phi_1 \in \text{range}\{\Delta\} = \text{range}\{d\delta\} + \text{range}\{\delta d\}$. Consequently,

$$C^\infty(\Lambda^p M) = \ker\{\Delta\} + \text{range}\{d\delta\} + \text{range}\{\delta d\}\,. \tag{5.3.a}$$

Let ϕ be a differential form. If $d\phi = 0$ and $\delta\phi = 0$, then clearly $\phi \in \ker\{\Delta\}$. Conversely, $\Delta\phi = 0$ implies

$$0 = (\Delta\phi, \phi)_{L^2} = ((d\delta + \delta d)\phi, \phi)_{L^2} = (\delta\phi, \delta\phi)_{L^2} + (d\phi, d\phi)_{L^2}\,.$$

Consequently,

$$d\phi = \delta\phi = 0 \quad\text{so}\quad \ker\{\Delta\} = \ker\{d\} \cap \ker\{\delta\}\,.$$

Expand $\phi = \Phi_0 + d\delta\Phi_2 + \delta d\Phi_3$. Then:

$$\begin{aligned}
(\Phi_0, \delta d\Phi_3)_{L_2} &= (d\Phi_0, d\Phi_3)_{L^2} = 0,\\
(\Phi_0, d\delta\Phi_2)_{L_2} &= (\delta\Phi_0, \delta\Phi_2)_{L^2} = 0,\\
(\delta d\Phi_3, d\delta\Phi_2)_{L^2} &= (\delta\delta d\Phi_3, \delta\Phi_2)_{L^2} = 0\,.
\end{aligned}$$

So Equation (5.3.a) is an orthogonal direct sum decomposition with respect to the L^2 inner product. If $d\phi = 0$, then $d\delta d\Phi_3 = 0$ so:

$$0 = (d\delta d\Phi_3, d\Phi_3)_{L^2} = (\delta d\Phi_3, \delta d\Phi_3)_{L^2}\,.$$

So $\delta d\Phi_3 = 0$ and, consequently, $\phi = \Phi_0 + d\delta\Phi_2$ and $[\phi] = [\Phi_0]$. If $\Phi_0 = d\Phi_1$ and $\Phi_0 \in \ker\{\Delta\}$, then we may show $\Phi_0 = 0$ and complete the proof by computing:

$$0 = (\delta\Phi_0, \Phi_1)_{L^2} = (\Phi_0, d\Phi_1)_{L^2} = (\Phi_0, \Phi_0)_{L^2}\,. \qquad \square$$

We apply the results of Lemma 5.8. Let ∇ be the Levi–Civita connection extended to act on tensors of all types. Let $\nabla_{\partial_{x^j}}\omega = \omega_{;j}$ and let $\nabla_{\partial_{x^i}}\nabla_{\partial_{x^j}}\omega = \omega_{;ji}$;

$$\nabla\omega = \omega_{;j}dx^j \quad\text{and}\quad \nabla^2\omega = \omega_{;ji} \otimes dx^i \otimes dx^j\,.$$

Let R_{ij}^{Λ} be the curvature operator of the Levi–Civita connection acting on the exterior algebra. We express the operators d and δ in terms of the Levi–Civita connection in Assertion 1, establish the Weitzenböch identity [61] in Assertion 2, and prove the Bochner Vanishing Theorem [5] in Assertion 3 of the following result; Assertion 1 played a crucial role in the proof of Theorem 4.9.

Theorem 5.14 *Let (M, g) be a compact connected Riemannian manifold without boundary. Let $\{e_i\}$ be a local orthonormal frame for TM and let $\{e^i\}$ be the associated dual frame field for T^*M. Let ω be a smooth differential form on M.*

1. $d\omega = \operatorname{ext}(e^i)\nabla_{e_i}\omega$ *and* $\delta\omega = -\operatorname{int}(e^i)\nabla_{e_i}\omega$.
2. $\Delta\omega = -g^{ij}\omega_{;ij} + \frac{1}{2}c(e^i)c(e^j)R^{\Lambda}_{ij}\omega$.
3. *If ω is a smooth 1-form, then* $\Delta^1\omega = -\operatorname{Tr}\{\nabla^2\omega\} + \operatorname{Ric}(\omega)$.
4. *If* $\operatorname{Ric} \geq 0$ *and if* $\operatorname{Ric} > 0$ *at a point of M, then* $H^1_{\mathrm{dR}}(M) = 0$.

Proof. Let $A := \operatorname{ext} \circ \nabla$ be a natural first order operator from $C^\infty(\Lambda^p M)$ to $C^\infty(\Lambda^{p+1} M)$. Relative to any system of local coordinates, we have

$$\begin{aligned} A(f_I dx^I) &= dx^i \wedge \nabla_{\partial_{x^i}}(f_I dx^I) = dx^i \wedge (\partial_{x^i} f_I) dx^I + dx^i \wedge f_I \nabla_{\partial_{x^i}} dx^I \\ &= d(f_I dx^I) + \mathcal{E}_1 \quad \text{where} \quad \mathcal{E}_1(f_I dx^I) := dx^i \wedge f_I \nabla_{\partial_{x^i}} dx^I . \end{aligned}$$

Since $\mathcal{E}_1 = A - d$, $\mathcal{E}_1$ is a natural 0th order partial differential operator (i.e., an endomorphism) which is linear in the Christoffel symbols. Fix a point P of M. By Lemma 3.13 in Book I, we can choose a system of local coordinates so the first derivatives of the metric vanish at P. Therefore, the Christoffel symbols vanish at P and $\mathcal{E}_1$ vanishes at P. Since P was arbitrary, $\mathcal{E}_1$ vanishes identically so $d = \operatorname{ext} \circ \nabla$. This shows that dually

$$\delta = \left\{\operatorname{ext}(dx^i)\nabla_{\partial_{x^i}}\right\}^* = -\operatorname{int}(dx^i)\nabla_{\partial_{x^i}} + \mathcal{E}_2 = -\operatorname{int} \circ \nabla + \mathcal{E}_2 ,$$

where $\mathcal{E}_2 := \delta - \operatorname{int} \circ \nabla$ is a natural 0th order partial differential operator which is linear in the first derivatives of the metric. The same argument shows $\mathcal{E}_2 = 0$ which establishes Assertion 1.

We use Assertion 1 to see $d + \delta = c \circ \nabla$ and, consequently,

$$\Delta = (d + \delta)^2\omega = c(dx^i)\nabla_{\partial_{x^i}} c(dx^j)\nabla_{\partial_{x^j}}\omega = c(dx^i)c(dx^j)\omega_{;ji} + \mathcal{E}_3 .$$

where again $\mathcal{E}_3$ is an invariantly defined 0th order operator which is linear in the 1-jets of the metric; hence again $\mathcal{E}_3 = 0$. We now establish Assertion 2 by computing:

$$\begin{aligned} (d + \delta)^2\omega &= \tfrac{1}{2}(c(dx^i)c(dx^j)\omega_{;ji} + c(dx^j)c(dx^i)\omega_{;ij}) \\ &= \tfrac{1}{2}(c(dx^i)c(dx^j) + c(dx^j)c(dx^i))\omega_{;ji} + \tfrac{1}{2}c(dx^j)c(dx^i)(\omega_{;ij} - \omega_{;ji}) \\ &= -g^{ij}\omega_{;ij} + \tfrac{1}{2}c(dx^j)c(dx^i)R(\partial_{x^j}, \partial_{x^i})\omega . \end{aligned}$$

We specialize to the case $p = 1$ and apply Assertion 2. Let $\{e_i\}$ be a local orthonormal frame for the tangent bundle and let $\{e^i\}$ be the dual frame. We may then lower and raise indices easily. We have $R(e_i, e_j)e^k = R_{ij}{}^k{}_\ell e^\ell$. Consequently,

$$c(e^i)c(e^j)R(e_i, e_j)e^k = R_{ij}{}^k{}_\ell c(e^i)c(e^j)c(e^\ell)1 . \tag{5.3.b}$$

Now if the indices $\{i, j, \ell\}$ are distinct, then

$$c(e^i)c(e^j)c(e^\ell) = c(e^j)c(e^\ell)c(e^i) = c(e^\ell)c(e^i)c(e^j)\,.$$

We have $R_{ij}{}^k{}_\ell = -R_{ij\ell}{}^k$. The Bianchi identity yields $R_{ij\ell}{}^k + R_{j\ell i}{}^k + R_{\ell ij}{}^k = 0$. Consequently, we may assume that the indices $\{i, j, \ell\}$ are not distinct in Equation (5.3.b). We cannot have $i = j$ as $R_{ij\ell}{}^k + R_{ji\ell}{}^k = 0$. Consequently, either $i = \ell$ or $j = \ell$; this yields the same thing. We suppose $i = \ell \neq j$. Then $c(e^i)c(e^j)c(e^\ell) = c(e^j)$. We may derive Assertion 3 from Assertion 2 by computing:

$$\tfrac{1}{2}c(e^i)c(e^j)R(e_i, e_j)e^k = R_{ij}{}^k{}_i c(e^j)1 = \rho_j{}^k e^j\,.$$

Suppose $\rho \geq 0$. Then

$$\begin{aligned}(\Delta^1\omega, \omega)_{L^2} &= (-\omega_{;ii}, \omega)_{L^2} + (\rho\omega, \omega)_{L^2}\\ &= (\nabla\omega, \nabla\omega)_{L^2} + (\rho\omega, \omega)_{L^2} \geq (\rho\omega, \omega)_{L^2} \geq 0\,.\end{aligned}$$

Consequently, if ω is a smooth 1-form with $\Delta^1\omega = 0$, then $\nabla\omega = 0$ so ω is parallel and hence $\|\omega\|$ is constant. But since $\rho > 0$ at some point, we have $(\rho\omega, \omega)(P) = 0$ so $\omega(P) = 0$. Since $\|\omega\|$ is constant, this implies ω vanishes identically which proves Assertion 4. □

5.3.3 POINCARÉ DUALITY. The following result is due to the French mathematician J. Poincaré in the topological setting.

Jules Henri Poincaré (1854–1912)

Let $\star$ be the Hodge operator and let $c(\text{dvol})$ be Clifford multiplication by the volume form as discussed in Section 5.2. We shall apply Lemma 5.9 and use the Hodge Decomposition Theorem to establish Poincaré duality [53]. If V and W are finite-dimensional vector spaces, then we say that a map f from $V \times W$ to $\mathbb{R}$ is a *perfect pairing* if f is bilinear, if given any v in V there exists w in W so that $f(v, w) \neq 0$, and if given any w in W there exists v in V so that $f(v, w) \neq 0$. Equivalently, f exhibits V as the dual of W and W as the dual of V. Let

$$\mathcal{I}(\omega_p, \theta_{m-p}) := \int_M \omega_p \wedge \theta_{m-p} \quad \text{for } \omega_p \in C^\infty(\Lambda^p M) \text{ and } \theta_{m-p} \in C^\infty(\Lambda^{m-p} M)\,. \tag{5.3.c}$$

Theorem 5.15 *Let M be a compact connected oriented Riemannian manifold without boundary.*

1. *The Hodge operator $\star$ defines an isomorphism from* $\ker\{\Delta^p\}$ *to* $\ker\{\Delta^{m-p}\}$.
2. *The map $\mathcal{I}$ defines a perfect pairing from $H^p_{\mathrm{dR}}(M) \times H^{m-p}_{\mathrm{dR}}(M)$ to $\mathbb{R}$.*
3. $\dim\{H^p_{\mathrm{dR}}(M)\} = \dim\{H^{m-p}_{\mathrm{dR}}(M)\}$.
4. $H^m_{\mathrm{dR}}(M) = [\mathrm{dvol}] \cdot \mathbb{R}$ *is generated by the volume form.*
5. *The map $[\omega_m] \to \int_M \omega_m$ is an isomorphism from $H^m_{\mathrm{dR}}(M)$ to $\mathbb{R}$.*

Proof. We use Stokes' Theorem to show the pairing of Equation (5.3.c) extends to cohomology. Suppose $d\omega_p = 0$ and $d\theta_{m-p} = 0$. If $\omega_p = d\psi_{p-1}$, then

$$d(\psi_{p-1} \wedge \theta_{m-p}) = \omega_p \wedge \theta_{m-p} + (-1)^{p-1}\psi_{p-1} \wedge d\theta_{m-p} = \omega_p \wedge \theta_{m-p},$$

$$\int_M \omega_p \wedge \theta_{m-p} = \int_M d(\psi_{p-1} \wedge \theta_{m-p}) = \int_{\partial M} \psi_{p-1} \wedge \theta_{m-p} = 0\,.$$

Consequently, the pairing extends to cohomology in the first factor. We use the identity $\mathcal{I}(\omega_p, \theta_{m-p}) = (-1)^{p(m-p)}\mathcal{I}(\theta_{m-p}, \omega_p)$ to see that the pairing also extends to cohomology in the second factor. By Theorem 5.14, $d + \delta = c(dx^i)\nabla_{\partial_{x^i}}$. Since $\xi * \mathrm{dvol} = (-1)^{m-1}\,\mathrm{dvol} * \xi$ in $\mathrm{Clif}(T^*M)$ for any $\xi \in T^*M$, $c(\xi)c(\mathrm{dvol}) = (-1)^{m-1}c(\mathrm{dvol})c(\xi)$. Consequently,

$$c(dx^i)\nabla_{\partial_{x^i}} c(\mathrm{dvol}) = (-1)^{m-1}c(\mathrm{dvol})c(dx^i)\nabla_{\partial_{x^i}} + \mathcal{E}\,,$$

where $\mathcal{E}$ is linear in the 1-jets of the metric and invariantly defined. Consequently, it vanishes since at any given point $P \in M$ we can always choose the coordinate system so that the first derivatives of the metric vanish at P by Lemma 3.13 in Book I. This proves that:

$$c(\mathrm{dvol})(d + \delta) = (-1)^{m-1}(d + \delta)c(\mathrm{dvol})\,.$$

Therefore, $c(\mathrm{dvol})\Delta = \Delta c(\mathrm{dvol})$ so $c(\mathrm{dvol}) : \ker\{\Delta^p\} \to \ker\{\Delta^{m-p}\}$. Since $c(\mathrm{dvol})^2 = \pm\,\mathrm{Id}$ and since $c(\mathrm{dvol}) = \pm\star$, Assertion 1 follows. Let $0 \neq \omega_p \in \ker\{\Delta^p\}$. We show that $\mathcal{I}$ is a perfect pairing by noting:

$$\mathcal{I}(\omega_p, \star\omega_p) = \int_M \omega_p \wedge \star\omega_p = \int_M g(\omega_p, \omega_p)|\,\mathrm{dvol}\,| = (\omega_p, \omega_p)_{L^2} \neq 0\,.$$

Assertion 3 follows from Assertion 1. Since M is connected, $H^0_{\mathrm{dR}}(M) = \mathbb{R} \cdot [1]$. Assertion 4 now follows from Assertion 1 since $\star 1 = \mathrm{dvol}$. Assertion 5 follows since $\mathcal{I}([1], \cdot)$ is an isomorphism from $H^m_{\mathrm{dR}}(M)$ to $\mathbb{R}$. □

5.3.4 THE KÜNNETH FORMULA [38]. This result together with Poincaré duality is an extremely powerful tool in examining the ring structure of de Rham cohomology. The following result was established by the German mathematician Künneth.

H. Künneth (1892–1975)

Theorem 5.16 *Let M and N be compact manifolds of dimensions m and n, respectively, and let π_1 and π_2 be projection on the first and second factors, respectively. Then $\pi_1^* \wedge \pi_2^*$ is an isomorphism from $H^*_{\mathrm{dR}}M \otimes H^*_{\mathrm{dR}}N$ to $H^*_{\mathrm{dR}}(M \times N)$.*

Proof. Let g_M and g_N be Riemannian metrics on M and on N, respectively. We take the product metric $g := g_M + g_N$ on $M \times N$. Let

$$\vec{x} = (x^1, \dots, x^m) \quad \text{and} \quad \vec{y} = (y^1, \dots, y^n)$$

be local coordinates on M and on N, respectively. We use Theorem 5.14 to compute:

$$(d+\delta)_{M\times N} = c_{M\times N}(dx^i)\nabla^{M\times N}_{\partial_{x^i}} + c_{M\times N}(dy^a)\nabla^{M\times N}_{\partial_{y^a}}\,.$$

The Christoffel symbols decouple so $\nabla^{M\times N}_{\partial_{x^i}} = \nabla^M_{\partial_{x^i}}$ and $\nabla^{M\times N}_{\partial_{y^a}} = \nabla^N_{\partial_{y^a}}$. We have

$$\begin{aligned} &c(dx^i)c(dy^a) + c(dy^a)c(dx^i) = -2g(dx^i, dy^a) = 0,\\ &\nabla^M_{\partial_{x^i}} c(dy^a) = c(dy^a)\nabla^M_{\partial_{x^i}}, \quad \text{and} \quad \nabla^N_{\partial_{y^a}} c(dx^i) = c(dx^i)\nabla^N_{\partial_{y^a}}\,. \end{aligned}$$

Therefore, we may compute:

$$\begin{aligned} \Delta_{M\times N} &= (d+\delta)^2_{M\times N}\\ &= \quad c(dx^i)\nabla^M_{\partial_{x^i}} c(dx^j)\nabla^M_{\partial_{x^j}} + c(dx^i)\nabla^M_{\partial_{x^i}} c(dy^a)\nabla^N_{\partial_{y^a}}\\ &\quad +c(dy^a)\nabla^N_{\partial_{y^a}} c(dx^i)\nabla^M_{\partial_{x^i}} + c(dy^a)\nabla^N_{\partial_{y^a}} c(dy^b)\nabla^N_{\partial_{y^b}}\\ &= \Delta_M + \Delta_N\\ &\quad +c(dx^i)c(dy^a)\nabla^M_{\partial_{x^i}}\nabla^N_{\partial_{y^a}} + c(dy^a)c(dx^i)\nabla^N_{\partial_{y^a}}\nabla^M_{\partial_{x^i}}\,. \end{aligned}$$

Since $R(\partial_{x^i}, \partial_{y^a}) = 0$, $\nabla^N_{\partial_{y^a}}\nabla^M_{\partial_{x^i}} = \nabla^M_{\partial_{x^i}}\nabla^N_{\partial_{y^a}}$. Consequently,

$$\begin{aligned} \Delta_{M\times N} &= \Delta_M + \Delta_N + (c(dx^i)c(dy^a) + c(dy^a)c(dx^i))\nabla^M_{\partial_{x^i}}\nabla^N_{\partial_{y^a}}\\ &= \Delta_M + \Delta_N\,. \end{aligned}$$

Let $\{\phi^M_{p,\nu}, \lambda^M_{p,\nu}\}$ be a complete spectral resolution of Δ^p_M and $\{\psi^N_{q,\mu}, \lambda^N_{q,\mu}\}$ be a complete spectral resolution of Δ^q_N. Suppose first $p = q = 0$. Let $\mathcal{A}$ be the subalgebra of $C^\infty(M \times N)$ generated by $C^\infty(M)$ and $C^\infty(N)$. By the Stone–Weierstrass Theorem, $\mathcal{A}$ is dense in

$$C^\infty(M \times N).$$

Since $\operatorname{span}\{\phi^M_{0,\nu} \cdot \psi^N_{0,\mu}\}$ is dense in $\mathcal{A}$, $\{\phi^M_{0,\nu} \cdot \psi^N_{0,\mu}\}$ is a complete orthonormal basis for $L^2(\Lambda^0(M \times N))$. A similar argument shows

$$\{\phi^M_{p,\nu} \wedge \psi^N_{q,\mu}\}_{p+q=r,\,1\le\nu<\infty,1\le\mu<\infty}$$

is a complete orthonormal basis for $L^2(\Lambda^r(M \times N)) = \oplus_{p+q=r} L^2(\Lambda^p M \wedge \Lambda^q N)$. This shows that

$$\{\phi^M_{p,\nu} \wedge \psi^N_{q,\mu}, \lambda^M_{p,\nu} + \lambda^N_{q,\mu}\}$$

is the complete spectral resolution of $\Delta^r_{M\times N}$ and

$$\ker\{\Delta^r_{M\times N}\} = \oplus_{p+q=r} \ker\{\Delta^p_M\} \wedge \ker\{\Delta^q_N\}.$$

Theorem 5.16 now follows from Theorem 5.13 by identifying de Rham cohomology with the kernel of the Laplacian. □

The ring structure is a more subtle invariant than just the additive structure. For example, the additive structure of the cohomology groups of the Cartesian product of $S^2 \times S^4$ and the additive structure of complex projective space $\mathbb{CP}^3$ is the same. We apply Theorem 5.4, Theorem 5.6, and Theorem 5.16 to see:

$$\begin{aligned}
H^0_{\mathrm{dR}}(S^2 \times S^4) &= [1] \times \mathbb{R}, & H^0_{\mathrm{dR}}(\mathbb{CP}^3) &= [1] \times \mathbb{R},\\
H^2_{\mathrm{dR}}(S^2 \times S^4) &= x_2 \times \mathbb{R}, & H^2_{\mathrm{dR}}(\mathbb{CP}^3) &= y_2 \times \mathbb{R},\\
H^4_{\mathrm{dR}}(S^2 \times S^4) &= x_4 \times \mathbb{R}, & H^4_{\mathrm{dR}}(\mathbb{CP}^3) &= y_4 \times \mathbb{R},\\
H^6_{\mathrm{dR}}(S^2 \times S^4) &= x_6 \times \mathbb{R}, & H^6_{\mathrm{dR}}(\mathbb{CP}^3) &= y_6 \times \mathbb{R}.
\end{aligned}$$

The ring structures, however, are completely different. By Theorem 5.16, the generators x_i of the de Rham cohomology of $S^2 \times S^4$ can be chosen so that $x_2^2 = 0$ and $x_6 = x_2 \wedge x_4$. On the other hand, we will show in Theorem 5.18 that the generators of the de Rham cohomology of $\mathbb{CP}^3$ can be chosen so $y_4 = y_2 \wedge y_2$ and $y_6 = y_2 \wedge y_4$. So the cohomology ring structures tell these two spaces apart. The cohomology ring structure will also play a crucial role in the determination of the de Rham cohomology of the unitary group $\mathrm{U}(n)$ in Theorem 6.30 where we will show

$$H^*_{\mathrm{dR}}(\mathrm{U}(n)) = \Lambda[x_1, x_3, \dots, x_{2n-1}]$$

is an exterior algebra on the Maurer–Cartan generators x_{2i-1} in degrees $2i - 1$ for $1 \le i \le n$.

5.4 CHARACTERISTIC CLASSES

5.4.1 THE PULLBACK BUNDLE. Let $\pi_N : V \to N$ be a real or complex vector bundle. If F is a smooth map from M to N, then the *pullback bundle* F^*V is defined by the *pullout diagram*

$$F^*V := \{(x, v) \in M \times V : F(x) = \pi_N(v)\}.$$

Projection on the first factor, $\pi_M(x, v) := x$, from F^*V to M makes F^*V into a vector bundle. If

$$\{\phi_{\alpha\beta} : \mathcal{O}_\alpha \cap \mathcal{O}_\beta \to \text{GL}(n, \cdot)\}$$

are the transition functions for V relative to an open cover $\mathcal{O}_\alpha$ of N, then

$$\{F^*\phi_{\alpha\beta} : F^{-1}(\mathcal{O}_\alpha) \cap F^{-1}(\mathcal{O}_\beta) \to \text{GL}(n, \cdot)\}$$

are the transition functions for F^*V. If s is a smooth section to V over N, the pullback section is defined by setting $(F^*s)(x) := (x, s(F(x)))$. If ∇ is a connection on V over N, then $F^*\nabla$ is the connection on F^*V characterized by the property that $(F^*\nabla)(F^*s) = F^*(\nabla s)$ for any smooth section s to V over N. The Christoffel symbols of the pullback connection are the pullback of the Christoffel symbols.

5.4.2 THE FIRST CHERN CLASS. S. Chern was a Chinese mathematician who is in many respects the father of characteristic class theory.

S. Chern (1911–2004)

Let $\text{Vect}^1_{\mathbb{C}}(M)$ be the set of equivalence classes of complex line bundles over M; $\text{Vect}^1_{\mathbb{C}}(M)$ is a group under tensor product. The first Chern class is a group homomorphism from $\text{Vect}^1_{\mathbb{C}}(M)$ to $H^2_{\text{dR}}(M)$ which is a natural transformation of functors that may be defined as follows. Let ∇ be a connection on a complex line bundle L over M. If s is a local non-zero section to L, let $\nabla(s) = \omega_s s$ where ω_s is the associated *connection 1-form*. Set $\Omega(\nabla, s) := d\omega_s$. If s_1 and s_2 are two local non-zero sections to L over an open set $\mathcal{O}$, then $s_1 = f s_2$ where f is a smooth function from $\mathcal{O}$ to $\mathbb{C} - \{0\}$. We then have

$$f\omega_{s_1} s_2 = \omega_{s_1} s_1 = \nabla s_1 = \nabla(f s_2) = df \cdot s_2 + f\omega_{s_2} s_2 .$$

Consequently, $f\omega_{s_1} = df + f\omega_{s_2}$ or, equivalently,

$$\omega_{s_1} = f^{-1} df + \omega_{s_2} . \tag{5.4.a}$$

Since $d\{f^{-1}df\} = 0$, $d\omega_{s_1} = d\omega_{s_2}$. Consequently, $\Omega(\nabla) := \Omega(\nabla, s) = d\omega_s$ is independent of the particular local section chosen and is globally defined. If $\tilde{\nabla}$ is another connection, then $(\nabla - \tilde{\nabla}) = \theta$ is a globally defined 1-form so $d\omega - d\tilde{\omega} = d\theta$ and $[\Omega(\nabla)] = [\Omega(\tilde{\nabla})]$ in de Rham cohomology. If ∇ is Hermitian and if $\|s\|^2 = 1$, then

$$0 = d\langle s, s\rangle = \langle \nabla s, s\rangle + \langle s, \nabla s\rangle = \omega_s + \bar{\omega}_s .$$

Consequently, $d\omega_s$ is purely imaginary and $\frac{\sqrt{-1}}{2\pi} d\omega_s$ is real. Let

$$c_1(L) := \left[\tfrac{\sqrt{-1}}{2\pi} d\omega_s\right] \in H^2_{\mathrm{dR}}(M)$$

define the *first Chern class*. Let $\mathbb{L}$ be the tautological line bundle defined in Equation (4.3.b).

Theorem 5.17 *The first Chern class has the following properties.*

1. $c_1(L_1 \otimes L_2) = c_1(L_1) + c_1(L_2)$.
2. *If* $f : M \to N$, *then* $f^* c_1(L_N) = c_1(f^* L_N)$.
3. *If* $\mathbb{L}$ *is the tautological line bundle over* $S^2 = \mathbb{CP}^1$, *then* $\int_{\mathbb{CP}^1} c_1(\mathbb{L}) = 1$.

Proof. If ∇_i are connections on L_i, we may define a connection ∇ on $L_1 \otimes L_2$ by setting

$$\nabla(s_1 \otimes s_2) = \nabla_1 s_1 \otimes s_2 + s_1 \otimes \nabla_2 s_2 .$$

We then have $\omega = \omega_1 + \omega_2$ and Assertion 1 follows. If ∇ is a connection on a line bundle over L with connection 1-form ω, then $f^*\nabla$ is a connection on f^*L with connection 1-form $f^*\omega$ and Assertion 2 follows. We will derive Assertion 3 from a more general result in Theorem 5.18 presently. □

One can use classifying space theory to see that Theorem 5.17 completely characterizes c_1; we omit the details as it is beyond the scope of our present investigation.

5.4.3 THE FIRST CHERN CLASS OF A HOLOMORPHIC LINE BUNDLE.

Let L be a holomorphic line bundle which is equipped with a Hermitian inner product $\langle \cdot, \cdot \rangle$ and which is defined over a holomorphic manifold M. Let ∂ and $\bar{\partial}$ be as defined in Section 4.3.2. If

$$z = (z^1, \ldots, z^m) \quad \text{for} \quad z^a = x^a + \sqrt{-1} y^a$$

are local holomorphic coordinates on M and if $f \in C^\infty(M)$, then

$$\begin{aligned} &\partial f := \partial_{z^i} f \cdot dz^i \quad \text{and} \quad \bar{\partial} f := \partial_{\bar{z}^i} f \cdot d\bar{z}^i \quad \text{where} \\ &\partial_{z^i} := \tfrac{1}{2}(\partial_{x^i} - \sqrt{-1}\partial_{y^i}) \quad \text{and} \quad \partial_{\bar{z}^i} := \tfrac{1}{2}(\partial_{x^i} + \sqrt{-1}\partial_{y^i}) . \end{aligned}$$

Let s be a local holomorphic section to L. Define $\omega_s = \partial\log\langle s, s\rangle$. If $\tilde{s}$ is another local holomorphic section to L, then we may express $s = f\tilde{s}$ where f is holomorphic. Then $\langle s, s\rangle = f\bar{f}\langle\tilde{s}, \tilde{s}\rangle$ and, consequently,

$$\omega_s = \partial\log\{f\bar{f}\langle\tilde{s}, \tilde{s}\rangle\} = \partial\log(f) + \partial\log(\bar{f}) + \partial\log(\langle\tilde{s}, \tilde{s}\rangle) = df\cdot f^{-1} + \omega_{\tilde{s}}\,.$$

This is the transition law given in Equation (5.4.a). Therefore, ∇ is a well-defined connection on L. Because $d = \partial + \bar{\partial}$ and since $\partial\partial = 0$, we have

$$c_1(\nabla) = \tfrac{\sqrt{-1}}{2\pi}d\partial\log\|s\|^2 = \tfrac{\sqrt{-1}}{2\pi}\bar{\partial}\partial\log\|s\|^2\,.$$

5.4.4 THE COHOMOLOGY RING STRUCTURE OF $\mathbb{CP}^m$. In Theorem 5.6, we determined the additive structure of the de Rham cohomology of complex projective space. We now use the first Chern class to determine the ring structure of $H^*_{\mathrm{dR}}(\mathbb{CP}^m)$.

Theorem 5.18 *Let $\mathbb{L}$ be the tautological line bundle over $\mathbb{CP}^m$.*

1. $\int_{\mathbb{CP}^m} c_1(\mathbb{L})^m = 1$.
2. $H^{2k}_{\mathrm{dR}}(\mathbb{CP}^m) = c_1(\mathbb{L})^k\cdot\mathbb{C}$ *for* $1 \le k \le m$.
3. $H^*_{\mathrm{dR}}(\mathbb{CP}^m) = \mathbb{C}[c_1(\mathbb{L})]/(c_1(\mathbb{L})^{m+1} = 0)$ *is a truncated polynomial ring.*

Proof. We recall the discussion of Section 4.3.3. Let π be the natural projection from $\mathbb{C}^{m+1} - \{0\}$ to $\mathbb{CP}^m$. Then $\mathcal{O}_i := \pi\{z \in \mathbb{C}^{m+1} : z^i \neq 0\} \subset \mathbb{CP}^m$ is an open cover of $\mathbb{CP}^m$ with associated local coordinates given by

$$\Theta_i(w^1, \dots, w^m) := \pi(w^1, \dots, 1, \dots, w^m)\,.$$

The *tautological line bundle* $\mathbb{L}$ is given by

$$\mathbb{L} := \{\xi \times z \in \mathbb{CP}^m \times \mathbb{C}^{m+1} : z \in \xi\}$$

and we have local sections $s_i(w)$ on $\mathcal{O}_i$ given in terms of the local parameter ω by

$$s_i(w) := \Theta_i(w) \times (w^1, \dots, 1, \dots, w^m) \in C^\infty(\mathbb{L}|_{\mathcal{O}_i})\,.$$

Let $\langle\cdot,\cdot\rangle$ be the standard inner product on $\mathbb{C}^{m+1}$; $\langle s_1(w), s_1(w)\rangle = 1 + |w|^2$. We shall restrict our integrals to $\mathcal{O}_1$ as $\mathbb{CP}^m - \mathcal{O}_1 = \mathbb{CP}^{m-1}$ has measure zero in $\mathbb{CP}^m$. We have

$$c_1(\mathbb{L}) = \tfrac{\sqrt{-1}}{2\pi}\bar\partial\partial\log(1+|w|^2) = \tfrac{\sqrt{-1}}{2\pi}\bar\partial\left\{\tfrac{\bar w^1 dw^1+\cdots+\bar w^m dw^m}{1+|w|^2}\right\}$$
$$= -\tfrac{\sqrt{-1}}{2\pi}\left\{(1+|w|^2)^{-1}\textstyle\sum_a dw^a\wedge d\bar w^a - (1+|w|^2)^{-2}\sum_{a,b}\bar w^a w^b dw^a\wedge d\bar w^b\right\}.$$

Note that $|\operatorname{dvol}|_e := dx^1\wedge dy^1\wedge\cdots\wedge dx^m\wedge dy^m$ is the Euclidean volume form on $\mathbb{C}^m$. We may then express $c_1(\mathbb{L})^m = f(w,\bar w)|\operatorname{dvol}|_e$. The construction is invariant under the unitary group. Consequently, $f(w,\bar w) = f(|w|)$ depends only on the distance from the origin. Note that $dw^a\wedge d\bar w^a = 2\sqrt{-1}dx^a\wedge dy^a$. We evaluate at the point $(r,0,\dots,0)$ to see

$$\begin{aligned} c_1(\mathbb{L})^m &= \pi^{-m}\left\{\sum_{a\ge 2}(1+r^2)^{-1}dx^a\wedge dy^a + (1+r^2)^{-2}dx^1\wedge dy^1\right\}^m \\ &= \pi^{-m}(1+r^2)^{-m-1}m!|\operatorname{dvol}|. \end{aligned}$$

This enables us to compute

$$\int_{\mathbb{CP}^m} c_1(\mathbb{L})^m = \int_{\mathcal{O}_1} c_1(\mathbb{L})^m = m!(\pi)^{-m}\operatorname{Vol}(S^{2m-1})\int_0^\infty (1+r^2)^{-m-1}r^{2m-1}dr.$$

We apply Lemma 3.3 of Book I to see that if m is even, then

$$\operatorname{vol}(S^{2m-1}) = \frac{(2\pi)^m}{2\cdot 4\cdot\cdots\cdot(2m-2)} = \frac{2\pi^m}{(m-1)!}.$$

Therefore, after changing variables to let $t = r^2$ and integrating by parts repeatedly, we have:

$$\begin{aligned} \int_{\mathbb{CP}^m} c_1(\mathbb{L})^m &= 2m\int_0^\infty r^{2m-1}(1+r^2)^{-m-1}dr = m\int_0^\infty t^{m-1}(1+t)^{-m-1}dt \\ &= (m-1)\int_0^\infty t^{m-2}(1+t)^{-m}dt = \cdots = \int_0^\infty (1+t)^{-2}dt = 1. \end{aligned}$$

This proves Assertion 1. Assertion 1 shows $c_1(\mathbb{L})^m$ is non-zero. Consequently,

$$c_1(\mathbb{L})^k \ne 0 \quad\text{in}\quad H^{2k}(\mathbb{CP}^m) \quad\text{for}\quad 1\le k\le m.$$

The additive structure of $\mathbb{CP}^m$ is given by Theorem 5.6. Consequently, we have that $c_1(\mathbb{L})^k$ is an additive generator of $H^{2k}_{\mathrm{dR}}(\mathbb{CP}^m)$ and $H^{2k}_{\mathrm{dR}}(\mathbb{CP}^m) = c_1(\mathbb{L})^k\cdot\mathbb{C}$. □

We shall not discuss the Chern classes more generally as this would take us too far afield and instead shall refer to Gilkey [20] for further information concerning the Chern classes noting that there are many excellent references on this subject.

CHAPTER 6

Lie Groups

In Chapter 6, we discuss Lie groups; the Norwegian mathematician Marius Sophus Lie introduced many of the basic concepts in this area.

S. Lie (1842–1899)

In Section 6.1, we give the basic definitions. In Section 6.2, we define the Lie algebra $\mathfrak{g}$ that is associated to a Lie group G. This correspondence is a very fruitful one since we can recover many of the geometric properties of a Lie group from the algebraic properties of its Lie algebra; $\mathfrak{g}$ in many ways can be regarded as the linearization of G. A good reference for this section is Kirillov [36]. In Section 6.3, we define the exponential map from $\mathfrak{g}$ to G which can be used to give local coordinates to G. We shall be primarily interested in compact Lie groups. We shall show in Theorem 6.18 that every compact Lie group is diffeomorphic to a closed matrix group. Therefore, we shall restrict our attention to matrix groups where it is possible to define the exponential and the log maps by power series. We will use the exponential map to introduce the orthogonal, special linear, and unitary groups. In Section 6.4, we show that any closed subgroup of a Lie group is a Lie group in its own right. We use this result to discuss the classical groups. In Section 6.5, we discuss the representation theory of compact Lie groups and prove the Peter–Weyl Theorem. In Section 6.6, we shall turn to Riemannian geometry. If g is a bi-invariant metric, we relate the exponential map of Riemannian geometry and the exponential map of a Lie group. In Section 6.7, we define the Killing form and discuss relationships between the Killing form and various algebraic and geometric properties of the Lie group in question. In Section 6.8, we examine the groups

$$\{\mathrm{SO}(3),\ S^3,\ \mathrm{SU}(2),\ \mathrm{SO}(4),\ S^3\times S^3,\ \mathrm{SL}(2,\mathbb{R})\}\,.$$

There are some relations among these groups that are important. In Section 6.9, we discuss the de Rham cohomology of a compact Lie group G. We express the de Rham cohomology of G in terms of its Lie algebra and show the de Rham cohomology of G is the exterior algebra on finitely many generators in odd degrees. In Section 6.10, we determine the de Rham cohomology of the unitary group.

6.1 BASIC CONCEPTS

If $\mathbb{F}$ denotes the real numbers, the complex numbers, or the quaternions, and if V and W are $\mathbb{F}$ vector spaces, let $\mathrm{Hom}_{\mathbb{F}}(V, W)$ be the set of all $\mathbb{F}$ linear maps from V to W, i.e., homomorphisms in the appropriate category; $\mathrm{Hom}_{\mathbb{F}}(V, W)$ is a $\mathbb{F}$ vector space. When $\mathbb{F}$ is clear by context, we shall simplify the notation and write $\mathrm{Hom}(V, W)$, and if $V = W$, we shall write $\mathrm{Hom}(V)$.

A smooth m-dimensional manifold G is said to be a *Lie group* if G is equipped with a group structure so that the multiplication map $(g_1, g_2) \to g_1 \cdot g_2$ and the inverse $g \to g^{-1}$ are smooth maps from $G \times G$ to G and from G to G, respectively. Let e be the identity element of G. We say that G is *Abelian* if the group multiplication is commutative. The following examples are instructive.

6.1.1 THE REAL FIELD $\mathbb{R}$. The half-line $(0, \infty)$ is an Abelian group with respect to multiplication. The map $x \to e^x$ provides an isomorphism between $(\mathbb{R}, +)$ and $((0, \infty), \cdot)$. More generally, any real m-dimensional vector space $(V, +)$ is an Abelian Lie group with respect to vector sum. The determinant is a well-defined map $\det : \mathrm{Hom}(V) \to \mathbb{R}$. An element T in $\mathrm{Hom}(V)$ is invertible if and only if $\det(T) \neq 0$. Let

$$\mathrm{GL}(V) := \{A \in \mathrm{Hom}(V) : \det(A) \neq 0\}$$

be the set of all invertible elements of $\mathrm{Hom}(V)$. This is an open subset of $\mathrm{Hom}(V)$ and hence is a smooth manifold. Since $\det(A_1 A_2) = \det(A_1)\det(A_2)$, $\mathrm{GL}(V)$ is closed under matrix product and becomes a group under composition. If we choose a basis to identify V with $\mathbb{R}^m$, then we may identify $\mathrm{GL}(V)$ with the group of invertible $m \times m$ real matrices, $\mathrm{GL}(m, \mathbb{R})$. Since matrix multiplication is polynomial, operator composition defines a smooth map from $\mathrm{GL}(V) \times \mathrm{GL}(V)$ to $\mathrm{GL}(V)$. Cramer's rule (see Lemma 1.3 in Book I) expresses the inverse in terms of determinants of minors and shows that the map $A \to A^{-1}$ is smooth. Thus, $\mathrm{GL}(V)$ is a Lie group. Although, as noted above, we can always identify $\mathrm{GL}(V)$ with $\mathrm{GL}(m, \mathbb{R})$, it is convenient occasionally to be able to work in a basis free fashion.

6.1.2 THE COMPLEX FIELD $\mathbb{C}$. Addition makes the complex field $\mathbb{C}$ into a Lie group; multiplication makes the non-zero complex numbers $\mathbb{C}^* := \mathbb{C} - \{0\}$ into a Lie group. The unit circle $S^1 := \{z \in \mathbb{C} : |z| = 1\}$ is then a Lie subgroup of $\mathbb{C}^*$. The map $z \to e^{\sqrt{-1}z}$ provides group homomorphisms from $(\mathbb{C}, +)$ to $(\mathbb{C}^*, \cdot)$ and from $(\mathbb{R}, +)$ to $(S^1, \cdot)$. These maps are covering projections but are not injective since $e^{2\pi\sqrt{-1}} = 1$. If V is a complex vector space, then the group of invertible complex linear transformations $\mathrm{GL}_{\mathbb{C}}(V)$ of V is a Lie group of real dimension $2m^2$ that can be identified with the group of invertible $m \times m$ complex matrices $\mathrm{GL}(m, \mathbb{C})$.

6.1.3 THE SKEW FIELD OF QUATERNIONS $\mathbb{H}$. Let $\mathbb{H}$ be the *quaternions* as discussed in Section 1.2.1 of Book I; $\mathbb{H}$ is a *skew-field.* Then $\mathbb{H}^* := \mathbb{H} - \{0\}$ is a Lie group under quaternion multiplication. Let $S^3 := \{z \in \mathbb{H} : |z| = 1\}$ be the unit sphere; quaternion multiplication restricts to give S^3 the structure of a Lie group.

6.1.4 PRODUCTS OF LIE GROUPS. If G_1 and G_2 are Lie groups, then the *Cartesian product* $G := G_1 \times G_2$ is a Lie group where we set $(g_1, g_2) \cdot (\tilde{g}_1, \tilde{g}_2) := (g_1\tilde{g}_1, g_2\tilde{g}_2)$. The unit and inverse are

$$e_G := (e_{G_1}, e_{G_2}) \quad \text{and} \quad (g_1, g_2)^{-1} = (g_1^{-1}, g_2^{-1}).$$

For example, the m-dimensional torus $\mathbb{T}^m = S^1 \times \cdots \times S^1$, which is the product of m copies of the circle, is a Lie group.

6.2 LIE ALGEBRAS

A *Lie algebra* over $\mathbb{F}$ is a vector space $\mathfrak{g}$ over $\mathbb{F}$ which is equipped with an $\mathbb{F}$ bilinear operator $[\cdot,\cdot]$ (which is called the *Lie bracket*) from $\mathfrak{g} \times \mathfrak{g}$ to $\mathfrak{g}$ satisfying the two properties:

$$\begin{aligned} &[x, y] = -[y, x] && \text{(skew-symmetry)},\\ &[[x, y], z] + [[y, z], x] + [[z, x], y] = 0 && \text{(the Jacobi identity)}. \end{aligned}$$

The *Jacobi identity* was introduced by the German mathematician Carl Jacobi.

C. Jacobi (1804–1851)

If $\mathfrak{h}$ is a $\mathbb{F}$ vector subspace of $\mathfrak{g}$ with $[\mathfrak{h}, \mathfrak{h}] \subset \mathfrak{h}$, then the restriction of $[\cdot,\cdot]$ to $\mathfrak{h}$ makes $\mathfrak{h}$ into a Lie algebra and $\mathfrak{h}$ is said to be a *Lie subalgebra* of $\mathfrak{g}$. If $[A, B] \in \mathfrak{h}$ for every $A \in \mathfrak{g}$ and every $B \in \mathfrak{h}$, then $\mathfrak{h}$ is said to be an *ideal of* $\mathfrak{g}$. The following examples are illustrative.

6.2.1 ABELIAN LIE ALGEBRAS. Let V be a $\mathbb{F}$ vector space. Define $[x, y] = 0$ for all $x, y \in V$. It is immediate that $[\cdot,\cdot]$ is skew-symmetric, bilinear, and satisfies the Jacobi identity. Such a structure is said to be an *Abelian Lie algebra.*

6.2.2 VECTOR FIELDS. If M is a smooth manifold, let $C^\infty(TM)$ be the vector space of all smooth vector fields on M. The *Lie bracket*, which was defined in Section 2.2.2 of Book I, is given by $[X, Y]f := XYf - YXf$ for $X, Y \in C^\infty(TM)$ and $f \in C^\infty(M)$. Clearly $[X, Y] = -[Y, X]$ is skew-symmetric. Furthermore, $(C^\infty(TM), [\cdot,\cdot])$ is an infinite-dimensional Lie algebra because the Jacobi identity is satisfied:

$$\begin{aligned} &\{[[X, Y], Z] + [[Y, Z], X] + [[Z, X], Y]\}f \\ &\quad = \quad XYZf - YXZf - ZXYf + ZYXf + YZXf - ZYXf \qquad (6.2.a)\\ &\qquad - XYZf + XZYf + ZXYf - XZYf - YZXf + YXZf = 0. \end{aligned}$$

6.2.3 MATRIX ALGEBRAS. Let V be a finite-dimensional $\mathbb{F}$ vector space. If A and B are linear maps from V to V, let $[A,B] := AB - BA$. This is bilinear and skew-symmetric. The Jacobi identity follows from Equation (6.2.a). Let $\mathfrak{gl}_{\mathbb{F}}(V) := (\operatorname{Hom}_{\mathbb{F}}(V), [\cdot,\cdot])$.

6.2.4 LIE SUBALGEBRAS OF MATRIX ALGEBRAS. Several *Lie subalgebras* of $\mathfrak{gl}_{\mathbb{F}}(V)$ will play a role in our subsequent discussion. Let

$$\mathfrak{sl}_{\mathbb{F}}(V) := \{A \in \mathfrak{gl}_{\mathbb{F}}(V) : \operatorname{Tr}\{A\} = 0\}\,.$$

As $\operatorname{Tr}\{AB - BA\} = 0$, $\mathfrak{sl}_{\mathbb{F}}(V)$ is an ideal of $\mathfrak{gl}_{\mathbb{F}}(V)$. If $\langle\cdot,\cdot\rangle$ is a bilinear form on V, let

$$\mathfrak{so}(V, \langle\cdot,\cdot\rangle) := \{A \in \operatorname{Hom}_{\mathbb{F}}(V) : \langle Av, w\rangle + \langle v, Aw\rangle = 0 \quad \forall\, v, w \in V\}\,.$$

If A and B belong to $\mathfrak{so}(V, \langle\cdot,\cdot\rangle)$, we show $[A,B]$ belongs to $\mathfrak{so}(V, \langle\cdot,\cdot\rangle)$ by computing:

$$\begin{aligned} \langle [A,B]v, w\rangle &= \langle (AB - BA)v, w\rangle = -\langle Bv, Aw\rangle + \langle Av, Bw\rangle \\ &= \langle v, BAw\rangle - \langle v, ABw\rangle = -\langle v, [A,B]w\rangle\,. \end{aligned}$$

For example, if $V = \mathbb{R}^m$ and if $\langle v, w\rangle = v_1 w_1 + \cdots + v_m w_m$, then

$$\mathfrak{so}(\mathbb{R}^m, \langle\cdot,\cdot\rangle) = \{A \in \mathfrak{gl}_{\mathbb{R}}(\mathbb{R}^m) : A + A^t = 0\}\,.$$

If $\mathbb{F} = \mathbb{C}$ and if $\langle\cdot,\cdot\rangle$ is a non-degenerate Hermitian inner product on V, then:

$$\mathfrak{u}(V, \langle\cdot,\cdot\rangle) := \{A \in \mathfrak{gl}_{\mathbb{C}}(V) : \langle Av, w\rangle + \langle v, Aw\rangle = 0 \quad \forall\, v, w \in V\}\,.$$

Let $\Omega(x,y) := \langle x, Jy\rangle$ be the Kähler form of a pseudo-Hermitian inner product space $(V, \langle\cdot,\cdot\rangle, J)$. The symplectic Lie algebra is given by

$$\mathfrak{sp}_{\mathbb{R}}(V, \langle\cdot,\cdot\rangle, J) := \{A \in \mathfrak{gl}_{\mathbb{R}}(V) : J^t A J = A\}\,.$$

6.2.5 THE CROSS PRODUCT. Let $\mathbb{R}^3$ be given the usual vector cross product $\times$. This is a skew-symmetric bilinear map with

$$e_1 \times e_2 = -e_2 \times e_1 = e_3, \quad e_2 \times e_3 = -e_3 \times e_2 = e_1, \quad e_3 \times e_1 = -e_1 \times e_3 = e_2\,.$$

It is easily verified directly that the Jacobi identity is satisfied. One can also give a less combinatorial argument. Let $\{i, j, k\}$ be the purely imaginary quaternions. Then

$$[i,j] = ij - ji = 2k, \quad [j,k] = jk - kj = 2i, \quad [k,i] = ki - ik = 2j\,.$$

We may identify $(\mathbb{R}^3, \times)$ with the Lie algebra of purely imaginary quaternions by setting

$$e_1 = i/\sqrt{2}, \qquad e_2 = j/\sqrt{2}, \qquad e_3 = k/\sqrt{2}\,.$$

Let $\langle\cdot,\cdot\rangle$ be the usual Euclidean inner product on $\mathbb{R}^3$. A basis for $\mathfrak{so}(\mathbb{R}^3,\langle\cdot,\cdot\rangle)$ is given by

$$A_1=\begin{pmatrix}0&1&0\\-1&0&0\\0&0&0\end{pmatrix},\quad A_2=\begin{pmatrix}0&0&1\\0&0&0\\-1&0&0\end{pmatrix},\quad A_3=\begin{pmatrix}0&0&0\\0&0&-1\\0&1&0\end{pmatrix}.$$

One computes $[A_1,A_2]=A_3$, $[A_2,A_3]=A_1$, and $[A_3,A_1]=A_2$. Therefore, $(\mathbb{R}^3,\times)$ is also isomorphic to $\mathfrak{so}(\mathbb{R}^3,\langle\cdot,\cdot\rangle)$. These isomorphisms will play an important role subsequently.

6.2.6 STRUCTURE CONSTANTS. Let $\{e_1,\dots,e_m\}$ be a basis for an m-dimensional $\mathbb{F}$ Lie algebra $(\mathfrak{g},[\cdot,\cdot])$. The *Lie algebra structure constants* $c_{ij}{}^k\in\mathbb{F}$ are defined by the relation:

$$[e_i,e_j]=c_{ij}{}^k e_k\,. \tag{6.2.b}$$

The fact that $[\cdot,\cdot]$ is skew-symmetric and satisfies the Jacobi identity then yields:

$$c_{ij}{}^k+c_{ji}{}^k=0\quad\text{and}\quad c_{ij}{}^r c_{rk}{}^s+c_{jk}{}^r c_{ri}{}^s+c_{ki}{}^r c_{rj}{}^s=0\,. \tag{6.2.c}$$

Conversely, given a collection of constants $c_{ij}{}^k\in\mathbb{F}$ satisfying Equation (6.2.c), we may define a Lie bracket by setting $[e_i,e_j]:=\sum_k c_{ij}{}^k e_k$. For example, if we consider the Lie algebra $(\mathbb{R}^3,\times)$, then the non-zero structure constants are given by

$$c_{12}{}^3=1,\quad c_{21}{}^3=-1,\quad c_{23}{}^1=1,\quad c_{32}{}^1=-1,\quad c_{31}{}^2=1,\quad c_{13}{}^2=-1\,.$$

6.2.7 LIE ALGEBRAS IN LOW DIMENSIONS. The Lie algebras are classified in dimension $m\le 5$ by Patera et al. [51]. Up to isomorphism, there are two 2-dimensional Lie algebras whose brackets are defined by

$$A_{2,0}:\quad [e_1,e_2]=0\quad\text{and}\quad A_{2,1}:\quad [e_1,e_2]=e_1\,.$$

These two Lie algebras can be extended trivially to 3-dimensional Lie algebras. In addition, there are nine families of 3-dimensional Lie algebras. Their non-zero brackets are given by:

$$\begin{array}{ll}
A_{3,1}: & [e_2,e_3]=e_1.\\
A_{3,2}: & [e_1,e_3]=e_1,\ [e_2,e_3]=e_1+e_2.\\
A_{3,3}: & [e_1,e_3]=e_1,\ [e_2,e_3]=e_3.\\
A_{3,4}: & [e_1,e_3]=e_1,\ [e_2,e_3]=-e_2.\\
A_{3,5}^a: & [e_1,e_2]=e_1,\ [e_2,e_3]=ae_2,\ 0<|a|<1.\\
A_{3,6}: & [e_1,e_3]=-e_2,\ [e_2,e_3]=e_1.\\
A_{3,7}^a: & [e_1,e_3]=ae_1-e_2,\ [e_2,e_3]=e_1+ae_2,\ 0<a.\\
A_{3,8}^a: & [e_1,e_3]=-2e_2,\ [e_1,e_2]=e_1,\ [e_2,e_3]=e_3.\\
A_{3,9}^a: & [e_1,e_2]=e_3,\ [e_2,e_3]=e_1,\ [e_3,e_1]=e_2.
\end{array}$$

The classification of 4-dimensional and 5-dimensional Lie algebras is more complicated.

6.2.8 THE LIE ALGEBRA OF A LIE GROUP. We shall associate to each m-dimensional Lie group G a real m-dimensional Lie algebra $\mathfrak{g}$; it is traditional to use capital Roman letters for the Lie group and lower case gothic letters for the corresponding Lie algebra. Many of the geometric properties of G are reflected by the algebraic properties of $\mathfrak{g}$. For example, a group homomorphism $\phi : G \to H$ gives rise to a corresponding Lie algebra homomorphism ϕ_* from $\mathfrak{g}$ to $\mathfrak{h}$. Furthermore, a simply connected Lie group G is determined by the corresponding Lie algebra $\mathfrak{g}$. We shall work with left-invariant structures for the most part; the map $g \to g^{-1}$ defines an anti-group homomorphism which is a diffeomorphism of G that interchanges left-invariant and right-invariant structures. Let L_g and R_g denote left and right multiplication by g. If $h \in G$, then L_g and R_g are the diffeomorphisms of G which are defined by

$$L_g : h \to g \cdot h \quad \text{and} \quad R_g : h \to h \cdot g\,.$$

Let $(L_g)_*$ be the *pushforward* which was discussed in Section 2.2.3 of Book I. If $f \in C^\infty(G)$ and if $X \in C^\infty(TG)$, then $\langle (L_g)_* X, f\rangle = \langle X, L_g^* f\rangle$. The set $\mathfrak{g}$ of *left-invariant vector fields* is defined by:

$$\mathfrak{g} := \mathfrak{g}(G) := \{X \in C^\infty(TM) : (L_g)_* X = X \text{ for all } g \in G\}\,.$$

This is a vector space over $\mathbb{R}$. Since L_g is a diffeomorphism of G, we have naturally that

$$[(L_g)_* X, (L_g)_* Y] = (L_g)_*[X, Y]\,.$$

Consequently, if X and Y are left-invariant, then $[X, Y]$ is left-invariant and the commutator of vector fields gives $\mathfrak{g}$ the structure of a Lie algebra.

Let $F : M \to \tilde{M}$ be a smooth map from a manifold M to a manifold $\tilde{M}$. If X and $\tilde{X}$ are vector fields on M and on $\tilde{M}$, respectively, then we say that $F_* X = \tilde{X}$ if

$$F_* X(P) = \tilde{X}(FP) \quad \text{for all} \quad P \in M\,.$$

We note that given a vector field X on M, there may not be a vector field $\tilde{X}$ on $\tilde{M}$ such that $F_* X = \tilde{X}$ since there may be two different points P_1 and P_2 with

$$F(P_1) = F(P_2) \quad \text{but} \quad F_* X(P_1) \neq F_* X(P_2)\,.$$

The following is a useful observation:

Lemma 6.1 Let $F : M \to \tilde{M}$, let $F_* X = \tilde{X}$, and let $F_* Y = \tilde{Y}$. Let Φ_t^X be the flow for X and $\Phi_t^{\tilde{X}}$ the flow for $\tilde{X}$. Then $F_*[X, Y] = [F_* X, F_* Y]$ and $F\Phi_t^X = \Phi_t^{\tilde{X}} F$.

Proof. Let $\tilde{f} \in C^\infty(\tilde{M})$ and let $P \in M$. We show $F_*[X, Y] = [F_* X, F_* Y]$ by computing:

$$\begin{aligned}
[F_* X, F_* Y]\{\tilde{f}\}(FP) &= \{F_* X(F_* Y(\tilde{f})) - F_* Y(F_* X(\tilde{f}))\}(FP) \\
&= \{X\{F^*(F_* Y(\tilde{f}))\} - Y\{F^*(F_* X(\tilde{f}))\}\}(P) \\
&= \{XY(F^*\tilde{f}) - YX(F^*\tilde{f})\}(P) = [X, Y](F^*\tilde{f})(P) \\
&= \{F_*[X, Y](\tilde{f})\}(P)\,.
\end{aligned}$$

Let $\gamma(t) = \Phi_t^X P$ and let $\tilde{\gamma}(t) = F(\gamma(t))$. We compute:

$$\partial_t \tilde{\gamma}(t) = F_* \dot{\gamma}(t) = F_* X(\gamma(t)) = \tilde{X}(F(\gamma(t))) = \tilde{X}(\tilde{\gamma}(t)).$$

Consequently, $\tilde{\gamma}$ is an integral curve for $\tilde{X}$ starting at $F(P)$ and $\Phi_t^{\tilde{X}}(FP) = F(\Phi_t^X)(P)$. □

We use Lemma 6.1 to establish the following result.

Lemma 6.2

1. Let $\mathfrak{g}$ be the Lie algebra of left-invariant vector fields on a Lie group G. The map $X \to X(e)$ defines a linear isomorphism which identifies $\mathfrak{g}$ with $T_e(G)$.

2. If $\xi \in T_e G$, let X be the left-invariant vector field with $X(e) = \xi$. Then the flow Φ_t^X exists for all time and $L_g \Phi_t^X = \Phi_t^X L_g$ for all $g \in G$ and all $t \in \mathbb{R}$.

3. Let $\exp^{\mathfrak{g}}(\xi) = \Phi_1^X(e)$. This defines a smooth map from $\mathfrak{g}$ to G so that the curves $\exp^{\mathfrak{g}}(t\xi)$ are integral curves starting at e for the vector field X. If we identify $\mathfrak{g}$ with $T_e G$, then $(\exp_*^{\mathfrak{g}})(0) = \mathrm{Id}$ and, consequently, $\exp^{\mathfrak{g}}$ is a diffeomorphism from a neighborhood of 0 in $\mathfrak{g}$ to a neighborhood of e in G. We let $\log^{\mathfrak{g}}$ be the inverse map.

4. Let F be a smooth group homomorphism from a Lie group G to a Lie group H. (Such a map is called a Lie group morphism). Then F_* defines a Lie algebra homomorphism $F_* : \mathfrak{g} \to \mathfrak{h}$. If F_* is the zero map on $\mathfrak{g}$ and if G is connected, then F is the trivial group homomorphism.

Proof. Let $X \in \mathfrak{g}$. Since $L_g(e) = g$ and $(L_g)_* X = X$, we have

$$X(h) = \{(L_g)_* X\}(h) = (L_g)_*(X(g^{-1}h)).$$

Setting $h = g$ then yields $X(g) = (L_g)_* X(e)$. Consequently, if $X(e) = 0$, $X(g) = 0$ for all g, then X vanishes identically. This shows that the map $X \to X(e)$ is injective. Conversely, given $\xi \in T_e(G)$, we set $X(g) := (L_g)_* \xi$. If $f \in C^\infty(G)$, then

$$(Xf)(g) = \{((L_g)_* \xi)(f)\}(g) = \xi(L_g^* f).$$

The function $F(g,h) = (L_g^* f)(h) = f(gh)$ is a smooth function of two variables. Consequently, $\xi F(g,h)|_{h=e}$ is a smooth function of g. This shows $X(g)$ is a smooth vector field. It is immediate that $X(e) = \xi$. We show that X is left-invariant and complete the proof of Assertion 1 by computing:

$$\begin{aligned}((L_h)_* X)(hg) &= (L_h)_* X(g) = (L_h)_*(L_g)_* X(e) = (L_h \circ L_g)_* X(e) \\ &= (L_{hg})_* X(e) = X(hg).\end{aligned}$$

The Fundamental Theorem of Ordinary Differential Equations is often called the Cauchy–Lipschitz Theorem or the Picard–Lindelöf Theorem and is named after Émile Picard, Ernst Lindelöf, Rudolf Lipschitz, and Augustin–Louis Cauchy. It deals with the existence and uniqueness of solutions to an ordinary differential equation.

A. Cauchy (1789–1857) R. Lipschitz (1832–1903) E. Picard (1856–1941)

For each $g \in G$, this result shows that there exists an open neighborhood $\mathcal{O}_g$ of g and there exists $\epsilon_g > 0$ so the flow Φ_t^X exists for $|t| \le \epsilon_g$ on $\mathcal{O}_g$. Since $(L_g)_* X = X$, L_g commutes with Φ_t^X, i.e., $L_g \Phi_t^X = \Phi_t^X L_g$. This shows that $\Phi_t^X = L_g \Phi_t^X L_{g^{-1}}$ is well-defined on $L_g(\mathcal{O}_e)$ for $|t| < \epsilon_e$. Consequently, we can choose ϵ uniformly on G. We use the semi-group property $\Phi_t^X \Phi_s^X = \Phi_{s+t}^X$ to extend the flow for all t and prove Assertion 2. Since $\Phi_{ts}^X = \Phi_t^{sX}$,

$$\exp^{\mathfrak{g}}(tX) = \Phi_1^{tX} = \Phi_t^X ;$$

Assertion 3 follows.

Let $F : G \to H$ be a group homomorphism. If $\xi \in T_{e_G} G$, choose $X \in \mathfrak{g}$ so $X(e_G) = \xi$. Let $\tilde{\xi} := F_* \xi \in T_{e_H} H$. Let $\tilde{X} \in \mathfrak{h}$ satisfy $\tilde{X}(e_H) = \tilde{\xi}$. Since $L_{Fh} F = F L_h$,

$$F_* X(h) = F_*(L_h)_* X(e) = (L_{F(h)})_* F_* X(e) = L_{F(h)} \tilde{\xi} = \tilde{X}(Fh)$$

and thus $F_* X = \tilde{X}$. Thus, by Lemma 6.1, F_* is a Lie algebra morphism. If $F_* = 0$, then

$$\partial_t F \exp^{\mathfrak{g}}(t\xi)|_{t=t_0} = X_{F_*\xi}^H (F \exp^{\mathfrak{g}}(t_0 \xi)) = 0$$

and, consequently, $F \exp^{\mathfrak{g}}(t\xi) = e_H$ for all t. Since $\exp^{\mathfrak{g}}$ is a local diffeomorphism from a neighborhood of 0 in $T_{e_G} G$ to G, this implies F is constant on a neighborhood of e_G. Since G is connected, F is constant; this shows Assertion 4. □

Suppose G is connected. We shall show in Corollary 6.20 that if G is compact, then $\exp^{\mathfrak{g}}$ is surjective. The exponential map need not be subjective if G is not compact; we will show in Lemma 6.25 that exponential map for $\mathrm{SL}(2, \mathbb{R})$ is not surjective.

6.2.9 THE ADJOINT REPRESENTATION. Let $\mathfrak{g}$ be the Lie algebra of a Lie group G. If $\xi \in \mathfrak{g}$, let $\mathrm{ad}(\xi) \in \mathrm{Hom}(\mathfrak{g})$ be defined by $\mathrm{ad}(\xi) : \eta \to [\eta, \xi]$. The Jacobi identity yields:

$$\begin{aligned} [\mathrm{ad}(\xi), \mathrm{ad}(\eta)]\gamma &= \mathrm{ad}(\xi)\,\mathrm{ad}(\eta)\gamma - \mathrm{ad}(\eta)\,\mathrm{ad}(\xi)\gamma = \mathrm{ad}(\xi)[\eta, \gamma] - \mathrm{ad}(\eta)[\xi, \gamma] \\ &= [\xi, [\eta, \gamma]] - [\eta, [\xi, \gamma]] = [[\xi, \eta], \gamma] = \mathrm{ad}[\xi, \eta]\gamma . \end{aligned}$$

Therefore, $\text{ad} : \mathfrak{g} \to \text{Hom}(\mathfrak{g})$ is a Lie algebra morphism. If $g \in G$, then $R_{g^{-1}}$ commutes with left multiplication and defines a representation $\text{Ad}(g)$ from G to $\text{Hom}(\mathfrak{g})$; this is also induced by the map $h \to ghg^{-1}$. We have $\text{Ad}_*(e) = \text{ad}$. More generally, let V be a finite-dimensional real vector space. A smooth map $\sigma : G \to \text{GL}(V)$ is a *representation* if σ is a group homomorphism; the differential $\sigma_* : \mathfrak{g} = T_e(G) \to \mathfrak{gl}_{\mathbb{F}}(V)$ is a Lie algebra homomorphism.

6.2.10 THE CAMPBELL–BAKER–HAUSDORFF FORMULA. The following result is due to Henry Frederick Baker, John Edward Campbell, and Felix Hausdorff.

H. Baker (1866–1956)

J. Campbell (1862–1924)

F. Hausdorff (1868–1942)

Theorem 6.3 *Let* $\Phi(z) := (z-1)^{-1}\ln(z) = 1 + \frac{1}{2}(1-z) + \frac{1}{3}(1-z)^2 + \dots$ *If* G *is a Lie group, then*

$$\begin{aligned}\log^{\mathfrak{g}}\{\exp^{\mathfrak{g}}(X)\exp^{\mathfrak{g}}(Y)\} &= Y + \int_0^1 \Phi(e^{t\,\text{ad}(X)}e^{\text{ad}(Y)})X dt \\ &= X + Y + \tfrac{1}{2}[X,Y] + \tfrac{1}{12}\{[X,[X,Y]] - [Y,[X,Y]]\} - \tfrac{1}{24}[Y,[X,[X,Y]]] + \cdots\end{aligned}$$

is given by a convergent power series with universal coefficients which are independent of the particular Lie group in question. One has that $\exp^{\mathfrak{g}}(tX) \cdot Y \cdot \exp^{\mathfrak{g}}(-tX) = \exp(\text{ad}(X))Y$.

We shall refer to [4, 6, 29] for the proof of Theorem 6.3 as is beyond the scope of our present discussion; a particularly agreeable modern reference is Miller [44] (pages 159–161), where one needs Ado's Theorem to pass from matrix Lie algebras to the general case. Theorem 6.3 implies:

Corollary 6.4 If G is a Lie group, then G has a natural real analytic structure. If Ψ is a Lie group morphism, then Ψ is real analytic.

Proof. We use $\exp^{\mathfrak{g}}$ to define local coordinates near 0. By Theorem 6.3, the group multiplication is real analytic in these coordinates near Id. Consequently, the local coordinates $L_g \circ \exp^{\mathfrak{g}}$ and $\log^{\mathfrak{g}} \circ L_{g^{-1}}$ defined near any element $g \in G$ give G the structure of a real analytic manifold. Let Ψ be a smooth group homomorphism from G to $\tilde{G}$ and let $\psi : \mathfrak{g} \to \tilde{\mathfrak{g}}$. Since ψ is linear, ψ is real analytic. Because we may express $\Psi = \exp^{\tilde{\mathfrak{g}}} \circ \psi \circ \log^{\mathfrak{g}}$ near e_G, Ψ is a real analytic map from a neighborhood of e_G in G to a neighborhood of $e_{\tilde{G}}$ in $\tilde{G}$. Since Ψ is a group homomorphism, we may express $\Psi = L_{\Psi(g)} \circ \Psi \circ L_{g^{-1}}$ and conclude that Ψ is real analytic near an arbitrary point of G. □

6.2.11 A NATURAL EQUIVALENCE OF FUNCTORS. The association of the Lie algebra to a Lie group provides an equivalence of categories between the category of simply connected Lie groups and finite-dimensional real Lie algebras. If G is a connected Lie group, then the universal cover $\tilde{G}$ is again a connected Lie group and we may identify the Lie algebras $\tilde{\mathfrak{g}} = \mathfrak{g}$. Note that if G is not connected, then there can be inequivalent group structures on the universal cover. For example, the universal cover of the orthogonal group $\mathrm{O}(m)$ has two different structures which are denoted by $\mathrm{Pin}^{\pm}$ for $m \geq 3$; this plays a crucial role in Mathematical Physics (see, for example, DeWitt-Morette and DeWitt [16]).

By Lemma 6.2, if $F : G \to H$ is a morphism of Lie groups, then there is a natural morphism $F_* : \mathfrak{g} \to \mathfrak{h}$. Clearly $(\mathrm{Id})_* = \mathrm{Id}$ and $(F_1 \circ F_2)_* = (F_1)_* \circ (F_2)_*$. This shows that $G \rightsquigarrow \mathfrak{g}$ is a *natural transformation of functors*. The *Cartan–Lie Theorem* is simply the assertion that this is an isomorphism. It was first established by E. Cartan – we refer to related work by S. Lie. There are many good proofs of this result; see, for example, Kirillov [36] and van Est [58].

E. Cartan (1869–1951)

Theorem 6.5

1. *If $\mathfrak{g}$ is a finite-dimensional real Lie algebra, then there exists a simply connected Lie group G (which is unique up to isomorphism) with $\mathfrak{g}(G)$ isomorphic to $\mathfrak{g}$.*

2. *Let G be a simply connected Lie group and let H be a Lie group. Let $\mathfrak{g}$ and $\mathfrak{h}$ be the associated Lie algebras. If $\mathfrak{f}$ is a Lie algebra morphism from $\mathfrak{g}$ to $\mathfrak{h}$, then there is a unique Lie group homomorphism $F : G \to H$ with $F_* = \mathfrak{f}$.*

A closely related result is *Ado's Theorem* [1] (see also the proofs given by Cartan [11] and Harish–Chandra [28]):

Theorem 6.6 (Ado). *Every finite-dimensional real Lie algebra is isomorphic to a matrix Lie algebra. Consequently, every Lie group is locally isomorphic to a matrix group.*

We will show in Lemma 6.25 that the universal cover of $\mathrm{SL}(2, \mathbb{R})$ is not a matrix group and admits no faithful finite-dimensional representation. A bit of care must be used with Theorem 6.6 since the matrix group in question may only be immersed in $\mathrm{GL}(m, \mathbb{R})$ and not properly immersed. For example, take $m = 4$ and consider

$$\mathfrak{g} := \begin{pmatrix} 0 & 1 & 0 & 0 \\ -1 & 0 & 0 & 0 \\ 0 & 0 & 0 & \sqrt{2} \\ 0 & 0 & -\sqrt{2} & 0 \end{pmatrix} \cdot \mathbb{R}\,.$$

Being 1-dimensional, $\mathfrak{g}$ is Abelian. Therefore, $\mathfrak{g}$ is a Lie subalgebra of $\mathfrak{gl}(4,\mathbb{R})$. The associated Lie group is $\mathbb{R}$ and parametrized by

$$G = \left\{ \begin{pmatrix} \cos(t) & \sin(t) & 0 & 0 \\ -\sin(t) & \cos(t) & 0 & 0 \\ 0 & 0 & \cos(\sqrt{2}t) & \sin(\sqrt{2}t) \\ 0 & 0 & -\sin(\sqrt{2}t) & \cos(\sqrt{2}t) \end{pmatrix} : t \in \mathbb{R} \right\}.$$

This is immersed but the immersion is not proper as G is not a closed subgroup of $\mathrm{GL}(4,\mathbb{R})$; the topology of G as a Lie group is not the topology induced by the inclusion of G in $\mathrm{GL}(4,\mathbb{R})$. The closure of G in $\mathrm{GL}(4,\mathbb{R})$ is a 2-dimensional torus parametrized by

$$\bar{G} = \left\{ \begin{pmatrix} \cos(\theta) & \sin(\theta) & 0 & 0 \\ -\sin(\theta) & \cos(\theta) & 0 & 0 \\ 0 & 0 & \cos(\phi) & \sin(\phi) \\ 0 & 0 & -\sin(\phi) & \cos(\phi) \end{pmatrix} : \theta, \phi \in [0, 2\pi] \right\}.$$

The Lie group G winds around in $\bar{G}$ with irrational slope and is a dense subgroup.

6.3 THE EXPONENTIAL FUNCTION OF A MATRIX GROUP

Of particular interest in our investigation are *matrix groups*, i.e., groups which are subgroups of $\mathrm{Hom}(V)$ for some finite-dimensional vector space over $\mathbb{F} \in \{\mathbb{R}, \mathbb{C}, \mathbb{H}\}$. Matters are much more concrete in that setting and a lot of extra formalism can be avoided. By Theorem 6.6, the local geometry of a Lie group always arises in this way.

In the setting of matrix groups, the exponential map is given by the usual exponential map. We proceed as follows. Let $(V, |\cdot|)$ be a normed vector space. We assume the field in question is $\mathbb{R}$ to simplify the discussion. Let $\|A\| := \max_{|v|=1} |v|$ be the operator norm defined in Equation (1.2.f) of Book I. If $A, B \in \mathrm{Hom}(V)$, then

$$\|AB\| \leq \|A\| \cdot \|B\| \quad \text{and} \quad \|A + B\| \leq \|A\| + \|B\|\,.$$

We wish to define $e^A = \sum_{n\geq 0} A^n/n!$. Let $s_j(A) = \mathrm{Id} + A + \cdots + \frac{1}{j!}A^j$ be the partial sum. If $k > j > 2\|A\|$, then

$$\begin{aligned}
\|s_k(A) - s_j(A)\| &= \|\tfrac{1}{(j+1)!}A^{j+1} + \cdots + \tfrac{1}{k!}A^k\| \\
&\leq \tfrac{1}{(j+1)!}\|A\|^{j+1} + \cdots + \tfrac{1}{k!}\|A\|^k \\
&\leq \tfrac{1}{(j+1)!}\|A\|^{j+1}\{1 + \tfrac{1}{j+2}\|A\| + \tfrac{1}{(j+2)(j+3)}\|A\|^2 + \cdots\} \\
&\leq \tfrac{1}{(j+1)!}\|A\|^{j+1}(1 + \tfrac{1}{2} + \tfrac{1}{4} + \cdots) \\
&\leq 2\tfrac{1}{(j+1)!}\|A\|^{j+1}.
\end{aligned}$$

This tends uniformly to zero as $j \to \infty$ and hence, by the *Cauchy criteria*, we may define:

$$e^A = \lim_{n\to\infty} s_n(A) = \sum_{n=0}^{\infty} \frac{A^n}{n!}. \tag{6.3.a}$$

Lemma 6.7 Adopt the notation established above.

1. If $AB = BA$, then $e^A e^B = e^{A+B}$.
2. $e^{tA}e^{sA} = e^{(t+s)A}$.
3. We may identify $T_{\mathrm{Id}}(\mathrm{GL}(V)) = \mathrm{Hom}(V)$. Under this identification, $\exp^{\mathfrak{g}}(A) = e^A$.
4. The map $A \to e^A$ is a smooth map from $\mathrm{Hom}(V)$ to $\mathrm{GL}(V)$ which is a local diffeomorphism from $(\mathrm{Hom}(V), 0)$ to $(\mathrm{GL}(V), \mathrm{Id})$; denote the inverse by log.
5. We have $\det(e^A) = e^{\mathrm{Tr}\{A\}}$ for any $A \in \mathrm{Hom}(V)$.
6. If $\|A\| < 1$, then $\log(\mathrm{Id} + A) = A - \frac{1}{2}A^2 + \frac{1}{3}A^3 + \cdots$.

Proof. All the series in question converge absolutely and, consequently, the terms can be rearranged to suit. If A and B commute, we establish Assertion 1 by using the Binomial Theorem:

$$(A+B)^n = \sum_{i+j=n} \frac{n!}{i!j!} A^i B^j,$$

$$e^A e^B = \sum_{i,j} \frac{A^i}{i!}\frac{B^j}{j!} = \sum_{n=0}^{\infty} \frac{1}{n!} \sum_{i+j=n} \frac{n!}{i!j!} A^i B^j = \sum_n \frac{(A+B)^n}{n!} = e^{A+B}.$$

Assertion 2 now follows from Assertion 1. Since $e^A e^{-A} = e^0 = \mathrm{Id}$, $e^A \in \mathrm{GL}(V)$. Let A belong to $\mathrm{Hom}(V) = T_{\mathrm{Id}}(\mathrm{GL}(V))$. Then $(L_g)_* A = gA$. Let X be the left-invariant vector field agreeing with A at the identity; $X(g) = gA$ for $g \in \mathrm{GL}(V)$. Let $\Phi(g, A, t) := ge^{tA}$. By Equation (6.3.a),

$$\partial_t \Phi(g, A, t) = ge^{tA}A = X(\Phi(g, A, t)).$$

Consequently, $\Phi(g, A, t)$ gives the flow for X and $\exp^{\mathfrak{g}}(A) = \Phi(\mathrm{Id}, A, 1) = e^A$ which establishes Assertion 3. The map $\exp^{\mathfrak{g}}$ is a smooth map from $T_{\mathrm{Id}}\,\mathrm{GL}(V) = \mathrm{Hom}(V)$ to $\mathrm{GL}(V)$ by the Fundamental Theorem of Ordinary Differential Equations. Therefore, $A \to e^A$ is a smooth map; this shows Assertion 4.

By replacing V by $V \otimes_{\mathbb{R}} \mathbb{C}$, we may assume that V is complex in proving Assertion 5 and Assertion 6. By choosing a basis for V, we may assume that $V = \mathbb{C}^m$. Suppose first that $A = \mathrm{diag}(\lambda_1, \dots, \lambda_m)$ is diagonal. Then $e^A = \mathrm{diag}(e^{\lambda_1}, \dots, e^{\lambda_m})$ and Assertion 5 holds. Since the determinant and the trace are independent of the basis chosen, Assertion 5 continues to hold if A is diagonalizable. Since the determinant, trace, and exponential function are continuous, Assertion 5 continues to hold if A can be uniformly approximated by diagonalizable matrices. We may use Jordan normal form to see that the diagonalizable matrices are dense in $\mathcal{M}_n(\mathbb{C})$; Assertion 5 now follows in complete generality.

We now prove Assertion 6. The power series in Assertion 6 is the usual power series for the log function. If A is diagonalizable, then the argument given to establish Assertion 5 shows there exists $\delta > 0$ so that $e^{\log(\mathrm{Id} + \epsilon A)} = \mathrm{Id} + \epsilon A$ for $0 \le \epsilon < \delta$. If $\|A\| < 1$, then both sides of this equation are well-defined for $0 \le \epsilon \le 1$ and real analytic in ϵ. It now follows using analytic continuation that $e^{\log(\mathrm{Id} + \epsilon A)} = \mathrm{Id} + \epsilon A$ for $\epsilon \in [0, 1]$ and, consequently, that $e^{\log(\mathrm{Id} + A)} = \mathrm{Id} + A$ if $\|A\| < 1$ and if A is diagonalizable. Again, since the diagonalizable matrices are dense in $\mathrm{Hom}(V)$ if V is complex, we conclude this identity holds in complete generality. This shows Assertion 6.
$\square$

6.3.1 COMMUTATOR OF FLOWS.

We can relate the commutator of the flows of two vector fields and their Lie bracket (see also Spivak [57], Ch. 5, Vol. 1).

Lemma 6.8 Let X and Y be smooth vector fields on a smooth manifold M. Let Φ_t^X and Φ_t^Y be the associated flows.

1. Let $X(\vec{x}) = a^1(\vec{x})\partial_{x^1} + \dots + a^m(\vec{x})\partial_{x^m} = a(\vec{x})$ in a system of local coordinates. Then
$$\Phi_t^X = \vec{x} + t a(\vec{x}) + \tfrac{1}{2} t^2 da(\vec{x}) \cdot a(\vec{x}) + O(t^3)\,.$$
2. Let $\tau(t; P) := \Phi^Y_{-\sqrt{t}}(\Phi^X_{-\sqrt{t}}(\Phi^Y_{\sqrt{t}}(\Phi^X_{\sqrt{t}}(P))))$. Then τ is differentiable at $t = 0$ and
$$\dot{\tau}(P)_{t=0} = [X, Y](P)\,.$$

Proof. Expand the flows in a Taylor series about $t = 0$:
$$\Phi_t^X(\vec{x}) = \vec{x} + t a_1(\vec{x}) + \tfrac{1}{2} t^2 a_2(\vec{x}) + O(t^3),$$
$$\Phi_t^Y(\vec{x}) = \vec{x} + t b_1(\vec{x}) + \tfrac{1}{2} t^2 b_2(\vec{x}) + O(t^3)\,.$$
Note that $f(\vec{x} + \Delta\vec{x}) = f(\vec{x}) + df(\vec{x})\Delta\vec{x} + O(\|\Delta\vec{x}\|^2)$. We expand:
$$\begin{aligned}
\sigma_t(\vec{x}) &:= \Phi^Y_{-t}\left\{\Phi^X_{-t}\left[\Phi^Y_t(\Phi^X_t(\vec{x}))\right]\right\}\\
&= \Phi^Y_{-t}\left\{\Phi^X_{-t}\left[\Phi^Y_t(\vec{x} + t a_1(\vec{x}) + \tfrac{1}{2} t^2 a_2(\vec{x}))\right]\right\} + O(t^3)\\
&= \Phi^Y_{-t}\left\{\Phi^X_{-t}\left[\vec{x} + t a_1(\vec{x}) + \tfrac{1}{2} t^2 a_2(\vec{x}) + t b_1(\vec{x} + t a_1(\vec{x})) + \tfrac{1}{2} t^2 b_2(\vec{x})\right]\right\} + O(t^3)\,.
\end{aligned}$$

Since $tb_1(\vec{x}+ta_1(\vec{x})) = tb_1(\vec{x}) + t^2 db_1(\vec{x})\cdot a_1(\vec{x}) + O(t^3)$, we see

$$\begin{aligned}\sigma_t(\vec{x}) &= \Phi^Y_{-t}\{\Phi^X_{-t}[\vec{x}+ta_1(\vec{x})+tb_1(\vec{x})\\ &\quad +t^2 db_1(\vec{x})\cdot a_1(\vec{x}) + \tfrac{1}{2}t^2 a_2(\vec{x}) + \tfrac{1}{2}t^2 b_2(\vec{x})]\} + O(t^3)\,.\end{aligned}$$

We continue the expansion:

$$\begin{aligned}\sigma_t(\vec{x}) &= \Phi^Y_{-t}\{\vec{x}+ta_1(\vec{x})+tb_1(\vec{x})+t^2 db_1(\vec{x})\cdot a_1(\vec{x}) + \tfrac{1}{2}t^2 a_2(\vec{x}) + \tfrac{1}{2}t^2 b_2(\vec{x})\\ &\quad -ta_1(\vec{x}+ta_1(\vec{x})+tb_1(\vec{x})) + \tfrac{1}{2}t^2 a_2(\vec{x})\} + O(t^3)\,.\end{aligned}$$

Since $-ta_1(\vec{x}+ta_1(\vec{x})+tb_1(\vec{x})) = -ta_1(\vec{x}) - t^2 da_1(\vec{x})\cdot a_1(\vec{x}) - t^2 da_1(\vec{x})\cdot b_1(\vec{x})) + O(t^3)$,

$$\begin{aligned}\sigma_t(\vec{x}) &= \Phi^Y_{-t}\{\vec{x}+tb_1(\vec{x})+t^2 db_1(\vec{x})\cdot a_1(\vec{x}) + t^2 a_2(\vec{x}) + \tfrac{1}{2}t^2 b_2(\vec{x})\\ &\quad -t^2 da_1(\vec{x})\cdot a_1(\vec{x}) - t^2 da_1(\vec{x})\cdot b_1(\vec{x}))\} + O(t^3)\,.\end{aligned}$$

Finally, we expand the action of Φ^Y_{-t} to compute

$$\begin{aligned}\sigma_t(\vec{x}) &= \vec{x}+tb_1(\vec{x})\\ &\quad +t^2 db_1(\vec{x})\cdot a_1(\vec{x}) + t^2 a_2(\vec{x}) + \tfrac{1}{2}t^2 b_2(\vec{x}) - t^2 da_1(\vec{x})\cdot a_1(\vec{x})\\ &\quad -t^2 da_1(\vec{x})\cdot b_1(\vec{x}) - tb_1(\vec{x}+tb_1(\vec{x})) + \tfrac{1}{2}t^2 b_2(\vec{x}) + O(t^3)\,.\end{aligned}$$

Again, we expand $-tb_1(\vec{x}+b_1(\vec{x})t) = -tb_1(\vec{x}) - t^2 db_1(\vec{x})\cdot b_1(\vec{x}) + O(t^3)$ to conclude:

$$\begin{aligned}\sigma_t(\vec{x}) &= \vec{x}+tb_1(\vec{x})+t^2 db_1(\vec{x})\cdot a_1(\vec{x}) + t^2 a_2(\vec{x}) + \tfrac{1}{2}t^2 b_2(\vec{x}) - t^2 da_1(\vec{x})\cdot a_1(\vec{x})\\ &\quad -t^2 da_1(\vec{x})\cdot b_1(\vec{x})) - tb_1(\vec{x}) - t^2 db_1(\vec{x})\cdot b_1(\vec{x}) + \tfrac{1}{2}t^2 b_2(\vec{x}) + O(t^3)\,.\end{aligned}$$

The coefficient of t is zero so this simplifies to become:

$$\begin{aligned}\sigma_t(\vec{x}) &= \vec{x}+t^2\{db_1(\vec{x})\cdot a_1(\vec{x}) - da_1(\vec{x})\cdot b_1(\vec{x})\}\\ &\quad +t^2 a_2(\vec{x}) - t^2 da_1(\vec{x})\cdot a_1(\vec{x}) + t^2 b_2(\vec{x}) - t^2 db_1(\vec{x})\cdot b_1(\vec{x}) + O(t^3)\,.\end{aligned} \tag{6.3.b}$$

Let $X(\vec{x}) = a(\vec{x})$ and $Y(\vec{x}) = b(\vec{x})$. The defining relation $\partial_t \Phi^X_t(\vec{x}) = X(\Phi_t(\vec{x}))$ yields:

$$a_1(\vec{x}) + ta_2(\vec{x}) + O(t^2) = a(\vec{x}+ta_1(\vec{x})) + O(t^2) = a(\vec{x}) + tda(\vec{x})\cdot a_1(\vec{x}) + O(t^2)\,.$$

Consequently, $a_1(\vec{x}) = a(\vec{x})$ and $a_2(\vec{x}) = da(\vec{x})\cdot a(\vec{x})$; Assertion 1 now follows. Similarly we have $b_1(\vec{x}) = b(\vec{x})$ and $b_2(\vec{x}) = db_1(\vec{x})\cdot b_1(\vec{x})$. Therefore, we may rewrite Equation (6.3.b) as:

$$\sigma_t(\vec{x}) = \vec{x} + t^2 db(\vec{x})\cdot a(\vec{x}) - t^2 da(\vec{x})\cdot b(\vec{x}) + O(t^3)\,. \tag{6.3.c}$$

Let $a = (a^1,\dots,a^m) = a^i e_i$ and let $b = (b^1,\dots,b^m) = b^j e_j$ where $\{e_i\}$ is the standard basis for $\mathbb{R}^m$. We replace 't' by '$\sqrt{t}$' to complete the proof by rewriting Equation (6.3.c) in the form $\sigma_{\sqrt{t}}(\vec{x}) = \vec{x} + te_i\cdot(a^j\partial_{x^j}b^i - b^j\partial_{x^j}a^i) + O(t^{3/2}) = \vec{x} + t\cdot[X,Y] + O(t^{3/2})$. □

6.3.2 LEFT-INVARIANT VECTOR FIELDS IN MATRIX GROUPS. If A belongs to $\operatorname{Hom}(V)$, let X_A^L (resp. X_A^R) be the left-invariant (resp. right-invariant) vector field which agrees with A at the identity. The following result relates the bracket on the Lie algebra of $\mathrm{GL}(V)$ with the ordinary matrix bracket and illustrates why we have chosen to work with left-invariant rather than with right-invariant vector fields.

Lemma 6.9 Let $A, B \in \operatorname{Hom}(V)$. Then $[X_A^L, X_B^L] = X_{[A,B]}^L$ and $[X_A^R, X_B^R] = -X_{[A,B]}^R$.

Proof. Let $\Phi_t^{X_A^L}$ and $\Phi_t^{X_B^L}$ be the flows. We use Lemma 6.7 and Lemma 6.8 to see:

$$\begin{aligned}
[X_A^L, X_B^L](\vec{x}) &= \partial_t|_{t=0}\left\{\Phi_{-\sqrt{t}}^{X_B^L}\Phi_{-\sqrt{t}}^{X_A^L}\Phi_{\sqrt{t}}^{X_B^L}\Phi_{\sqrt{t}}^{X_A^L}\right\}(\vec{x})\\
&= \partial_t|_{t=0}\left\{\vec{x}\cdot\exp^{\mathfrak{g}}(\sqrt{t}A)\exp^{\mathfrak{g}}(\sqrt{t}B)\exp^{\mathfrak{g}}(-\sqrt{t}A)\exp^{\mathfrak{g}}(-\sqrt{t}B)\right\}\\
&= \partial_t|_{t=0}\{\vec{x}\cdot[1+(AB-BA)t+O(t^{3/2})]\} = \vec{x}\cdot[A,B] = X_{[A,B]}^L(\vec{x}),\\
[X_A^R, X_B^R](\vec{x}) &= \partial_t|_{t=0}\left\{\Phi_{-\sqrt{t}}^{X_B^R}\Phi_{-\sqrt{t}}^{X_A^R}\Phi_{\sqrt{t}}^{X_B^R}\Phi_{\sqrt{t}}^{X_A^R}\right\}(\vec{x})\\
&= \partial_t|_{t=0}\left\{\exp^{\mathfrak{g}}(-\sqrt{t}B)\exp^{\mathfrak{g}}(-\sqrt{t}A)\exp^{\mathfrak{g}}(\sqrt{t}B)\exp^{\mathfrak{g}}(\sqrt{t}A)\cdot\vec{x}\right\}\\
&= \partial_t|_{t=0}\{[1+(BA-AB)t+O(t^{3/2})]\cdot\vec{x}\} = -[A,B]\cdot\vec{x} = -X_{[A,B]}^R(\vec{x}).
\end{aligned}$$

□

6.3.3 CLOSED SUBGROUPS OF LIE GROUPS. Let G be a Lie group. A submanifold H of G is said to be a *Lie subgroup* if H is also a subgroup of G. Note that H is necessarily closed in this setting. The converse is given by the following result.

Theorem 6.10 *A closed subgroup H of a Lie group G is a Lie subgroup.*

Proof. We follow the discussion in Hall [26] (pages 75–77). We shall assume that the ambient group is $\mathrm{GL}(V)$ as this simplifies the computations; the general case can be proved using exactly the same arguments although with a bit more technical fuss; one can also use Theorem 6.6 in this regard. To simplify the notation, we let $\exp := \exp^{\mathfrak{g}}$ be given by Equation (6.3.a) and let $\log := \log^{\mathfrak{g}}$ be the local inverse throughout the proof. We set

$$\mathfrak{h} := \{A \in M_m(\mathbb{R}) : \exp(tA) \in H \text{ for all } t\}.$$

Step 1. We first show $\mathfrak{h}$ is a linear subspace of $\operatorname{Hom}(V)$. If $A \in \mathfrak{h}$, then $sA \in \mathfrak{h}$ for all $s \in \mathbb{R}$ since $\exp(t(sA)) = \exp((ts)A)$. Let $A, B \in \mathfrak{h}$. We compute:

$$\begin{aligned}
&\log(\exp(tA)\exp(tB)) = \log((1+tA)(1+tB)+O(t^2)),\\
&\log(1+t(A+B)+O(t^2)) = t(A+B)+O(t^2),\\
&\log\left(\exp\left(\tfrac{t}{n}A\right)\exp\left(\tfrac{t}{n}B\right)\right)^n = n\log\left(\exp\left(\tfrac{t}{n}A\right)\exp\left(\tfrac{t}{n}B\right)\right),\\
&n\left(\tfrac{t}{n}(A+B)+O\left(\tfrac{t^2}{n^2}\right)\right) = t(A+B)+O\left(\tfrac{t^2}{n}\right),\\
&\exp(t(A+B)) = \lim_{n\to\infty}\left(\exp\left(\tfrac{t}{n}A\right)\exp\left(\tfrac{t}{n}B\right)\right)^n.
\end{aligned}$$

Consequently, $\exp(t(A+B))$ is the limit of elements of H and belongs to H as H is closed. This shows that $A+B \in \mathfrak{h}$.

Step 2. We show next that $[\mathfrak{h},\mathfrak{h}] \subset \mathfrak{h}$. Let $A, B \in \mathfrak{h}$. We have

$$\begin{aligned}
&\log\left\{\exp(tA)\exp(tB)\exp(-tA)\exp(-tB)\right\} \\
&\quad = \log\{(1+tA+\tfrac{t^2}{2}A^2)(1+tB+\tfrac{t^2}{2}B^2)(1-tA+\tfrac{t^2}{2}A^2) \\
&\qquad \times(1-tB+\tfrac{t^2}{2}B^2)+O(t^3)\} \\
&\quad = \log\{1+0\cdot t+(AB-BA)t^2+O(t^3)\} = t^2[A,B]+O(t^3),
\end{aligned}$$

$$t^2[A,B] = \lim_{n\to\infty} \log\left(\exp(\tfrac{t}{n}A)\exp(\tfrac{t}{n}B)\exp(-\tfrac{t}{n}A)\exp(-\tfrac{t}{n}B)\right)^{n^2},$$

$$\exp\left(t^2[A,B]\right) = \lim_{n\to\infty}\left(\exp\left(\tfrac{t}{n}A\right)\ \exp\left(\tfrac{t}{n}B\right)\exp\left(-\tfrac{t}{n}A\right)\exp\left(-\tfrac{t}{n}B\right)\right)^{n^2}.$$

Therefore, $\exp(t^2[A,B]) \in H$ for $t \geq 0$. Similarly, $\exp(-t^2[A,B]) = \exp(t^2[B,A]) \in H$ for any $t \geq 0$. Therefore, $[A,B] \in \mathfrak{h}$.

Step 3. We now show that there exists $\varepsilon > 0$ so that exp is a local homeomorphism from $B_\varepsilon(0) \cap \mathfrak{h}$ to a neighborhood of Id in H. Let $\mathfrak{h}^\perp$ be the complementary subspace. The map

$$(A, A^\perp) \to \exp(A)\exp(A^\perp)$$

is a local diffeomorphism from a neighborhood of 0 in $M_m(\mathbb{R})$ to a neighborhood of Id in $\mathrm{GL}(m,\mathbb{R})$. We suppose the theorem fails and argue for a contradiction. Then there exists a sequence of points $h_n \to \mathrm{Id}$ with $h_n \in H$ and $h_n \notin \exp(\mathfrak{h})$. Express

$$h_n = \exp(A_n)\exp(A_n^\perp).$$

Taking $\tilde{h}_n := \exp(-A_n)h_n = \exp(A_n^\perp)$, we see $\exp(A_n^\perp) \in H$ and $A_n^\perp \to 0$. Passing to a subsequence, we may assume $A_n^\perp = a_n\tilde{A}_n^\perp$ where $\tilde{A}_n^\perp$ are unit vectors with $\tilde{A}_n^\perp \to \tilde{A}^\perp$ and where $a_n \to 0$. Let $[\cdot]$ be the greatest integer function and let $j_n := [\frac{t}{a_n}]$. Then

$$\begin{aligned}
&\exp\left(a_n\tilde{A}_n^\perp\right) \in H \quad \Rightarrow \quad \exp\left(a_n\tilde{A}_n^\perp\right)^{j_n} \in H \quad \Rightarrow \quad \exp\left(j_na_n\tilde{A}_n^\perp\right) \in H, \\
&\exp(t\tilde{A}^\perp) = \exp(\lim_{n\to\infty} j_na_n\tilde{A}_n^\perp) = \lim_{n\to\infty}\exp\left(j_na_n\tilde{A}_n^\perp\right).
\end{aligned}$$

This shows $\tilde{A}^\perp \in \mathfrak{h}$. But $\tilde{A}^\perp \notin \mathfrak{h}$. Thus H is a smooth submanifold of $\mathrm{GL}(m,\mathbb{R})$. □

6.4 THE CLASSICAL GROUPS

We now use Theorem 6.10 to discuss the classical Lie groups (the orthogonal groups, the unitary groups, and the symplectic groups). We refer to Hall [26] and Helgason [30, 31] for further details. Our ambient Lie group will be $\mathrm{GL}(V)$ where V is a real, complex, or quaternion vector space with associated Lie algebra $\mathfrak{gl}_{\mathbb{F}}(V) := \mathrm{Hom}(V)$ with (by Lemma 6.9) the usual commutator bracket. These groups will all be closed subgroups of the general group and hence Lie subgroups.

6.4.1 THE GENERAL LINEAR GROUP. We continue the discussion of Section 1.2.4 of Book I. Let $\mathbb{F} \in \{\mathbb{R}, \mathbb{C}, \mathbb{H}\}$; $\mathrm{GL}(V)$ is a Lie group with Lie algebra $\mathfrak{gl}_{\mathbb{F}}(V) = \mathrm{Hom}(V)$.

6.4.2 THE SPECIAL LINEAR GROUP. $\mathrm{SL}(V) := \{g \in \mathrm{GL}(V) : \det(g) = 1\}$ over the field $\mathbb{F} = \mathbb{R}$ or $\mathbb{F} = \mathbb{C}$. It is necessary to assume $\mathbb{F} \neq \mathbb{H}$ as the quaternions are non-commutative and hence the determinant is not defined. By Lemma 6.7, $\det(e^A) = e^{\mathrm{Tr}\{A\}}$; the associated Lie algebra (see Section 6.2.4) is given by:

$$\mathfrak{sl}(V) := \{A \in \mathrm{Hom}(V) : \mathrm{Tr}\{A\} = 0\}.$$

6.4.3 THE ORTHOGONAL GROUP. Let $\mathrm{O}(V, \langle\cdot,\cdot\rangle) := \{g \in \mathrm{GL}(V) : g^*\langle\cdot,\cdot\rangle = \langle\cdot,\cdot\rangle\}$ where $\langle\cdot,\cdot\rangle$ is a non-degenerate real inner product of signature (p, q) on a real vector space V. Let A^* be the adjoint with respect to $\langle\cdot,\cdot\rangle$. The discussion of Section 6.2.4 shows that the Lie algebra is

$$\mathfrak{o}(V, \langle\cdot,\cdot\rangle) := \{A \in \mathrm{Hom}(V) : A + A^* = 0\}.$$

6.4.4 THE UNITARY GROUP. Let $\mathrm{U}(V, \langle\cdot,\cdot\rangle) := \{g \in \mathrm{GL}(V) : g^*\langle\cdot,\cdot\rangle = \langle\cdot,\cdot\rangle\}$ where $\langle\cdot,\cdot\rangle$ is a Hermitian symmetric inner product on a complex vector space of signature $(2p, 2q)$. If A^* is the adjoint, then the associated Lie algebra is

$$\mathfrak{u}(V, \langle\cdot,\cdot\rangle) := \{A \in \mathrm{Hom}(V) : A + A^* = 0\}.$$

Let $\mathrm{SU}(V, \langle\cdot,\cdot\rangle) := \mathrm{U}(V, \langle\cdot,\cdot\rangle) \cap \mathrm{SL}(V)$ be the special unitary group; the Lie algebra is given by $\mathfrak{su}(V, \langle\cdot,\cdot\rangle) := \{A \in \mathrm{Hom}(V) : A + A^* = 0 \text{ and } \mathrm{Tr}\{A\} = 0\}$.

6.4.5 THE SYMPLECTIC GROUP. Let $\mathrm{Sp}_{\mathbb{R}}(V, \langle\cdot,\cdot\rangle, J) := \{g \in \mathrm{GL}_{\mathbb{R}}(V) : g^*\Omega = \Omega\}$ where $\Omega(x, y) := \langle x, Jy\rangle$ is the Kähler form of a pseudo-Hermitian inner product space $(V, \langle\cdot,\cdot\rangle, J)$. The associated Lie algebra is

$$\mathfrak{sp}_{\mathbb{R}}(V, \langle\cdot,\cdot\rangle, J) := \{A \in \mathrm{Hom}_{\mathbb{R}}(V) : J^t A J = A\}.$$

6.4.6 SUMMARY. We collect in the following table the basic information about dimension, compactness and connectedness of the classical Lie groups. We take $V = \mathbb{R}^m$ or $\mathbb{C}^m$ and we take a positive definite orthogonal or unitary metric.

Group	Dimension	Compact	Connected	Components
$\mathrm{GL}(m,\mathbb{R})$	m^2	✗	✗	2
$\mathrm{GL}(m,\mathbb{C})$	$2m^2$	✗	✓	1
$\mathrm{SL}(m,\mathbb{R})$	m^2-1	✗	✓	1
$\mathrm{SL}(m,\mathbb{C})$	$2m^2-2$	✗	✓	1
$\mathrm{U}(m)$	m^2	✓	✓	1
$\mathrm{SU}(m)$	m^2-1	✓	✓	1
$\mathrm{O}(m)$	$m(m-1)/2$	✓	✗	2
$\mathrm{SO}(m)$	$m(m-1)/2$	✓	✓	1
$\mathrm{O}(m,\mathbb{C})$	$m(m-1)$	✗	✗	2
$\mathrm{SO}(m,\mathbb{C})$	$m(m-1)$	✗	✓	1
$\mathrm{Sp}(m,\mathbb{R})$	$m(m+1)/2$	✗	✓	1

The following result is a useful observation.

Lemma 6.11 $\mathrm{GL}(m)$ is homotopy equivalent to $\mathrm{O}(m)$ and $\mathrm{SL}(m,\mathbb{R})$ is homotopy equivalent to $\mathrm{SO}(m)$.

Proof. Let $A \in \mathrm{GL}(m,\mathbb{R})$. By applying the Gram–Schmidt process to the rows of A, we may use matrix multiplication to express $A = B \cdot C$ where $B \in T(m)$ is a lower triangular matrix with positive entries on the diagonal and where $C \in \mathrm{O}(m)$. We may express $B = \operatorname{diag}(e^{\epsilon_1},\dots,e^{\epsilon_m}) + \tilde{B}$ where $\tilde{B}$ is a strictly lower triangular matrix. Let $\phi_t(B) = \operatorname{diag}(e^{t\epsilon_1},\dots,e^{t\epsilon_m}) + t\tilde{B}$ define a deformation retract of $T(m)$ to the identity. This shows that $T(m)$ is contractible and, consequently, $\mathrm{GL}(m,\mathbb{R})$ is homotopy equivalent to $\mathrm{O}(m)$. If $A \in \mathrm{SL}(m,\mathbb{R})$, then $1 = \det(B)\cdot\det(C)$. Since $\det(B) > 0$, necessarily $\det(C) = 1$ so $C \in \mathrm{SO}(m)$. Therefore, $\mathrm{SL}(m,\mathbb{R})$ is homotopy equivalent to $\mathrm{SO}(m)$. □

6.5 REPRESENTATIONS OF A COMPACT LIE GROUP

Throughout this section we work over the complex field; the situation over the reals is very different. Let V be a complex vector space and let G be a compact Lie group. A *representation* of G on V is a smooth group homomorphism $\sigma : G \to \mathrm{GL}_{\mathbb{C}}(V)$; (V,σ) is a said to be a G*-module*. Let

$$g \cdot v := \sigma(g)v \quad \text{and} \quad V^G := \{v \in V : g \cdot v = v \text{ for all } g \in G\}.$$

We say that V is *irreducible* if the only G-invariant subspaces of V are $\{0\}$ and V. If W is a subspace of V which is invariant under G, then $(W, \sigma|_W)$ is said to be a *submodule*. If V_1 and V_2 are G-modules, let

$$\mathrm{Hom}^G(V_1,V_2) := \{T \in \mathrm{Hom}(V_1,V_2) : T\sigma_{V_1} = \sigma_{V_2}T\};$$

V and W are isomorphic G-modules if there exists $T \in \mathrm{Hom}^G(V,W)$ so T is bijective.

6.5.1 UNITIARIZING THE REPRESENTATION. We say that a smooth measure $d\mu$ on G is *bi-invariant* if it is invariant under both left and right translation, i.e., if

$$L_g^* d\mu = R_g^* d\mu = d\mu \quad \text{for all} \quad g \in G\,.$$

Lemma 6.12 Let G be a compact Lie group. Let (V, σ) be a G-module.

1. There is a unique smooth bi-invariant measure $|\,\mathrm{dvol}\,|$ on G with $\int_G |\,\mathrm{dvol}\,|(g) = 1$.
2. G admits a bi-invariant Riemannian metric whose volume element is $|\,\mathrm{dvol}\,|$.
3. There exists a Hermitian inner product $(\cdot, \cdot)$ on V so that $\sigma : G \to \mathrm{U}(V, (\cdot, \cdot))$.

Proof. Let $\{X_1, \dots, X_m\}$ be a basis for the Lie algebra $\mathfrak{g}$ of left-invariant vector fields on G and let $\{\omega^1, \dots, \omega^m\}$ be the corresponding dual basis for the left-invariant 1-forms. Let

$$\omega := \omega^1 \wedge \dots \wedge \omega^m\,.$$

Clearly, ω is left-invariant and any left-invariant m-form is a constant multiple of ω. Let $|\omega|$ be the associated measure. Since left and right multiplication commute, $R_h^*|\omega| = \chi(h)|\omega|$ for $\chi(h) \in \mathbb{R}^+$. Since $R_{h_1 h_2} = R_{h_2} R_{h_1}$, we have $\chi(h_1 h_2) = \chi(h_1)\chi(h_2)$. Consequently, χ is a representation from G to $\mathbb{R}^+$. Since G is compact, range$\{\chi\}$ is bounded. Since χ is a group homomorphism, range$\{\chi\}$ is a subgroup of $\mathbb{R}^+$. Consequently, $\chi \equiv 1$ and the measure $|\omega|$ is bi-invariant. By rescaling ω, we may assume $\int_G |\omega| = 1$. We set $|\,\mathrm{dvol}\,| := |\omega|$ to prove Assertion 1; the uniqueness is immediate from the construction.

Let g_0 be an arbitrary positive definite inner product on $\mathfrak{g}$. We extend g_0 to a left-invariant metric on T. If we average g_0 over the right action of g on $\mathfrak{g}$, we obtain a bi-invariant metric. This metric may be rescaled to ensure the associated volume form has total volume 1 and agrees with the measure defined in Assertion 1. This proves Assertion 2.

Let $\langle \cdot, \cdot \rangle$ be an arbitrary Hermitian inner product on V. We set

$$(v, w) = \int_G \langle \sigma(g)v, \sigma(g)w \rangle |\,\mathrm{dvol}\,|(g)\,.$$

We show that $\sigma(h)$ preserves $(\cdot, \cdot)$ for any $h \in G$ and establish Assertion 2 by computing:

$$\begin{aligned}
(\sigma(h)v, \sigma(h)w) &= \int_G \langle \sigma(g)\sigma(h)v, \sigma(g)\sigma(h)w \rangle |\,\mathrm{dvol}\,|(g) \\
&= \int_G \langle \sigma(gh)v, \sigma(gh)w \rangle |\,\mathrm{dvol}\,|(g) = \int_G \langle \sigma(gh)v, \sigma(gh)w \rangle |\,\mathrm{dvol}\,|(gh) \\
&= \int_G \langle \sigma(\tilde{g})v, \sigma(\tilde{g})w \rangle |\,\mathrm{dvol}\,|(\tilde{g}) = (v, w)\,.
\end{aligned}$$

□

6.5.2 EXAMPLE. Lemma 6.12 can fail if G is non-compact. By Lemma 6.23, the $ax+b$ group does not admit a smooth bi-invariant measure and the canonical representation of the $ax+b$ group on $\mathbb{R}^2$ does not admit an invariant orthogonal inner product. As an other example, we can let $G=\mathbb{R}$ and define a representation of $\mathbb{R}$ on $\mathbb{C}^2$ by

$$\sigma(x) := \begin{pmatrix} 1 & x \\ 0 & 1 \end{pmatrix}. \tag{6.5.a}$$

This operator has non-trivial Jordan normal form for $x \neq 0$ and is conjugate to no element of the unitary group. Consequently, V admits no invariant positive definite inner product.

6.5.3 DECOMPOSING REPRESENTATIONS INTO IRREDUCIBLES.

Lemma 6.13 Let $|\,\mathrm{dvol}\,|$ be the unique smooth bi-invariant measure of total mass 1 on a compact Lie group G. Let V and W be complex G-modules which are equipped with invariant Hermitian inner products.

1. Let π be orthogonal projection on V^G. Then $\pi = \int_G \sigma(g)|\,\mathrm{dvol}\,|(g)$.
2. If V and W are irreducible G-modules, then
$$\dim\{\mathrm{Hom}^G(V,W)\} = \left\{\begin{array}{ll} 1 & \text{if } V \text{ is isomorphic to } W \\ 0 \text{ if } V & \text{is not isomorphic to } W \end{array}\right\}.$$
3. Let W be a G-submodule of V. Then $W^\perp$ is a G-submodule of V.
4. There is an orthogonal direct sum decomposition $V = W_1 \oplus \cdots \oplus W_\ell$ of V into non-trivial irreducible submodules. If W is any irreducible G-module, let $n(V,W)$ be the number of the summands W_i which are isomorphic to W. Then $n(V,W) = \dim\{\mathrm{Hom}^G(W,V)\}$ is independent of the particular decomposition chosen.

Proof. If $v \in V^G$ is a fixed vector, then $\sigma(g)v = v$ for all g in G. Therefore,

$$\pi v = \int_G \sigma(g)v|\,\mathrm{dvol}\,|(g) = \int_G v|\,\mathrm{dvol}\,|(g) = v\,.$$

This shows $\pi = \mathrm{Id}$ on V^G. We show $\pi(v) \in V^G$ for any $v \in V$ by computing:

$$\begin{aligned}\sigma(h)\pi v &= \int_G \sigma(h)\sigma(g)v|\,\mathrm{dvol}\,|(g) = \int_G \sigma(hg)v|\,\mathrm{dvol}\,|(g)\\ &= \int_G \sigma(g)v|\,\mathrm{dvol}\,|(h^{-1}g) = \int_G \sigma(g)v|\,\mathrm{dvol}\,|(g) = \pi v\,.\end{aligned}$$

This shows range$\{\pi\} = V^G$. Since π is the identity on V^G, $\pi^2 = \pi$. Since $\langle\cdot,\cdot\rangle$ is an inner product invariant under the action of G, $\sigma(g)^* = \sigma(g)^{-1}$. We see that π is self-adjoint and establish Assertion 1 by computing:

$$\begin{aligned}
\pi^* v &= \int_G \sigma^*(g)v|\,\mathrm{dvol}\,|(g) = \int_G \sigma(g)^{-1}v|\,\mathrm{dvol}\,|(g) \\
&= \int_G \sigma(g^{-1})v|\,\mathrm{dvol}\,|(g) = \int_G \sigma(g)v|\,\mathrm{dvol}\,|(g^{-1}) \\
&= \int_G \sigma(g)v|\,\mathrm{dvol}\,|(g) = \pi v\,.
\end{aligned}$$

Let V be irreducible. Let $T \in \mathrm{Hom}^G(V)$ and let λ be a complex eigenvalue of T. Set

$$E_\lambda := \{v \in V : Tv = \lambda v\}\,.$$

If $v \in E_\lambda$ and if $g \in G$, then $T(g\cdot v) = g\cdot Tv = \lambda g\cdot v$ so E_λ is a non-trivial G-invariant subspace. Since V is irreducible, we conclude $E_\lambda = V$ and $T = \lambda\cdot\mathrm{Id}$. Consequently, $\dim\{\mathrm{Hom}^G(V)\} = 1$; it now follows $\dim\{\mathrm{Hom}^G(V,W)\} = 1$ if W is isomorphic to V. Conversely, suppose that $\dim\{\mathrm{Hom}^G(V,W)\} \neq 0$. Choose T non-zero in $\mathrm{Hom}^G(V,W)$. Since T is non-zero and $T(g\cdot v) = g\cdot Tv$, range$\{T\} \neq \{0\}$. Consequently, since W is irreducible, range$\{T\} = W$. Since $\ker\{T\} \neq V$ and V is irreducible, $\ker\{T\} = \{0\}$. Consequently, T provides a G isomorphism from V to W. This proves Assertion 2.

Let W be a G-invariant subspace of V. Let $w \in W$, let $w^\perp \in W^\perp$, and let $g \in G$. Since $\sigma(g^{-1})w \in W$, $(\sigma(g^{-1})w, w^\perp) = 0$. We show $\sigma(g)w^\perp \in W^\perp$ and establish Assertion 3:

$$(w, \sigma(g)w^\perp) = (\sigma(g^{-1})w, \sigma(g^{-1})\sigma(g)w^\perp) = (\sigma(g^{-1})w, w^\perp) = 0\,.$$

By applying Assertion 3 recursively, we can construct an orthogonal direct sum decomposition $V = W_1 \oplus \cdots \oplus W_\ell$ of V into irreducible modules W_i. We complete the proof of Assertion 4 by computing:

$$\begin{aligned}
\dim\{\mathrm{Hom}^G(W,V)\} &= \dim\{\mathrm{Hom}^G(W,W_1) \oplus \cdots \oplus \mathrm{Hom}^G(W,W_\ell)\} \\
&= \dim\{\mathrm{Hom}^G(W,W_1)\} + \cdots + \dim\{\mathrm{Hom}^G(W,W_\ell)\} \\
&= n(V,W)\,.
\end{aligned}$$

□

Assertion 2 of Lemma 6.13 fails over $\mathbb{R}$. Let $G = S^1$. Let

$$T(\theta) := \begin{pmatrix} \cos\theta & \sin\theta \\ -\sin\theta & \cos\theta \end{pmatrix}$$

define an action of S^1 on $\mathbb{R}^2$ by rotations. Since the only invariant subspaces are $\mathbb{R}^2$ and $\{0\}$, this representation is irreducible. However, $\mathrm{Id} \in \mathrm{Hom}^{S^1}(\mathbb{R}^2,\mathbb{R}^2)$ and, since G is Abelian, $T(\theta)$

belongs to $\text{Hom}^{S^1}(\mathbb{R}^2, \mathbb{R}^2)$ for any θ. This shows $\dim\{\text{Hom}^{S^1}(\mathbb{R}^2, \mathbb{R}^2)\} \geq 2$. What happens, of course, is that $T(\theta)$ does not have any eigenspaces for $\theta \in (0, \pi)$ in contrast to the complex setting.

Assertion 4 of Lemma 6.13 fails if G is not compact. Let $\sigma : \mathbb{R} \to \text{GL}_{\mathbb{C}}(2)$ be defined by Equation (6.5.a). The invariant subspaces of σ are $\{0\}$, $V_1 := \mathbb{C} \cdot (1, 0)$, and $\mathbb{C}^2$. There is no complementary invariant subspace to V_1 and we cannot decompose $\mathbb{C}^2 = V_1 \oplus V_2$ as the direct sum of irreducibles; σ exhibits non-trivial Jordan normal form and is not unitarizable in this instance.

6.5.4 THE ORTHOGONALITY RELATIONS. Let G be a compact Lie group. Let $L^2(G)$ be defined by the bi-invariant smooth measure of total volume 1 which is given by Lemma 6.12. If V is a complex G-module, let $(\cdot, \cdot)$ be a positive definite G-invariant inner product on V given by Lemma 6.12. Choose an orthonormal basis $\{e_i\}$ for V and let $\sigma(g)e_i = \xi^j_{V,i}(g)e_j$ be the matrix coefficients.

Lemma 6.14 Let G be a compact Lie group. Let V and W be irreducible complex G-modules.

1. If V and W are inequivalent, then $(\xi^v_{W,u}, \xi^j_{V,i})_{L^2(G)} = 0$ for all i, j, u, v.

2. $(\xi^v_{V,u}, \xi^j_{V,i})_{L^2(G)} = \frac{1}{\dim\{V\}} \delta_{ui} \delta_{vj}$ for all i, j, u, v.

Proof. We identify $\text{Hom}(V, W) = V^* \otimes W$ to define a natural G action on $\text{Hom}(V, W)$ so that $g(T) = \sigma_W(g) \circ T \circ \sigma_V(g)^*$; the natural inner product on $\text{Hom}(V, W)$ is then G-invariant as well. Because $\sigma_V(g)^* = \sigma_V(g^{-1})$, $g(T) = gTg^{-1}$. Consequently, if T belongs to $\text{Hom}^G(V, W)$, then $gT = Tg$. Fix i and v. We use Lemma 6.13 to express:

$$\begin{aligned}(\pi T)^v_i &= \int_G \xi^v_{W,u}(g) \cdot T^u_j \cdot \xi^j_{V,i}(g)^* |\,\text{dvol}\,|(g) \\ &= T^u_j \int_G \xi^v_{W,u}(g) \bar{\xi}^i_{V,j}(g) |\,\text{dvol}\,|(g) = T^u_j (\xi^v_{W,u}, \xi^i_{V,j})_{L^2(G)} \quad \text{for all} \quad i, v\,.\end{aligned} \tag{6.5.b}$$

If V and W are irreducible and inequivalent, then $\text{Hom}^G(V, W) = \{0\}$ by Lemma 6.13. Consequently, the orthogonal projection π from $\text{Hom}(V, W)$ to $\text{Hom}^G(V, W)$ is the zero map. Thus, $(\pi T)^v_i = 0$ so $0 = T^u_j (\xi^v_{W,u}, \xi^i_{V,j})_{L^2(G)}$. As T^u_j was arbitrary, $(\xi^v_{W,u}, \xi^i_{V,j})_{L^2(G)} = 0$ for all j, u as well. This proves Assertion 1.

Let $\dim\{V\} = r$. If $V = W$, then $\text{Hom}^G(V, V) = \mathbb{C} \cdot \text{Id}$ by Lemma 6.13. We take a basis $\{A_{ju}\}$ for $\text{Hom}(V)$ by setting $(A_{ju})^v_i := \delta^i_j \delta^v_u$. We have $A_{ju} : e_j \to e_u$. This is the matrix whose only non-zero entry is in position (u, j). The $\{A_{ju}\}$ are an orthonormal basis for $\text{Hom}(V)$. Since $\text{Id} = \sum_u A_{uu}$, $\|\,\text{Id}\,\|^2 = r$. Since A_{ab} has a 1 in position (a, b) and is zero elsewhere, $\langle A_{ab}, \text{Id} \rangle$ is 1 if $a = b$ and 0 otherwise. We use Equation (6.5.b) to establish Assertion 2 by computing:

$$r^{-1} \delta_{ab} \delta^v_i = r^{-1} \langle A_{ab}, \text{Id} \rangle\, \text{Id}^v_i = \pi(A_{ab})^v_i = (A_{ab})^u_j (\xi^v_{V,u}, \xi^i_{V,j})_{L^2(G)} = (\xi^v_{V,b}, \xi^i_{V,a})_{L^2(G)}\,. \quad \square$$

6.5.5 THE PETER–WEYL THEOREM. The left regular action $(L_g f)(h) := f(gh)$ makes $L^2(G)$ into a representation space G; $L^2(G)$ is finite-dimensional if and only if G is a finite group. Let $\mathrm{Irr}_{\mathbb{C}}(G)$ be the set of equivalence classes of irreducible complex modules for the group G. By Lemma 6.14, $\{\xi^j_{V,i}\}_{V\in\mathrm{Irr}_{\mathbb{C}}(G)}$ is an orthogonal subset of $L^2(G)$. Consequently, in particular, all these functions are linearly independent. If $V \in \mathrm{Irr}_{\mathbb{C}}(G)$, let:

$$A^j_V := \mathrm{span}_{1\le i\le\dim\{\xi\}}\{\xi^j_{V,i}\} \subset L^2(G) \text{ and } A_V := \mathrm{span}_{1\le i,j\le\dim\{\xi\}}\{\xi^j_{V,i}\} \subset L^2(G)\,.$$

Lemma 6.15 Let $V, W \in \mathrm{Irr}_{\mathbb{C}}(G)$.

1. $\dim\{A^j_V\} = \dim\{V\}$ and $\dim\{A_V\} = \dim\{V\}^2$.
2. $A^j_V \perp A^i_V$ in $L^2(G)$ for $i \neq j$.
3. $A_V \perp A_W$ in $L^2(G)$ for V not isomorphic to W.
4. $L_g A^j_V = A^j_V$ so A^j_V is a finite representation space for G.
5. A^j_V is isomorphic to V as a representation space for G.
6. Let $\tilde{A}$ be a finite-dimensional G-invariant subspace of $L^2(G)$ which is abstractly isomorphic to V as a representation space. Then $\tilde{A} \subset A_V$ in $L^2(G)$.
7. $\{\xi^j_{V,i}\}_{\xi\in\mathrm{Irr}_{\mathbb{C}}(G)}$ is a complete orthogonal basis for $L^2(G)$.
8. $L^2(G) = \oplus_{\xi\in\mathrm{Irr}_{\mathbb{C}}(G)} \oplus_{1\le j\le\dim\{\xi\}} A^j_\xi$.

Proof. The first three assertions follow from the orthogonality relations of Lemma 6.14. We now establish Assertion 4 and Assertion 5. We have that

$$\{L_g\xi^j_{V,i}\}(h) = \xi^j_{V,i}(gh) = \xi^\ell_{V,i}(g)\xi^j_{V,\ell}(h)\,.$$

This means we have the functional identity $L_g\xi^j_{V,i} = \xi^\ell_{V,i}(g)\xi^j_{V,\ell}$. Consequently, the space A^j_ξ is invariant under L_g and the corresponding matrix representation is given by $\xi^j_{V,i}$ relative to the canonical basis. Finally, suppose $\tilde{A}$ is a subspace of $L^2(G)$ whose dimension is finite. Also assume $\tilde{A}$ is abstractly isomorphic to V_ξ as a representation space. Choose a basis f_i for $\tilde{A}$ so that $L_g f_\nu = \xi^\mu_{V,\nu}(g) f_\mu$. Evaluating at the unit of the group shows

$$f_\nu(g) = (L_g f_\nu)(e) = \textstyle\sum_\mu \xi^\mu_{V,\nu}(g) f_\mu(e)\,.$$

Therefore, f_ν is a linear combination of the $\xi^\mu_{V,\nu}$ so $\tilde{A} \subset A$. This proves Assertion 6.

We use Theorem 5.11 to decompose $L^2(G) = \oplus_\lambda E(\lambda)$ into the eigenspaces of the scalar Laplacian. Each eigenspace $E(\lambda)$ is a representation space for G whose dimension is finite. We decompose each $E(\lambda)$ as the direct sum of irreducible modules. Each irreducible module is a subspace of some A_ξ and, consequently, $E(\lambda) \subset \oplus_\xi A_\xi$. This shows $L^2(G) \subset \oplus_\xi A_\xi$; the reverse inclusion is trivial. □

The following result, called the Peter–Weyl Theorem [52], was proved by Hermann Weyl and his student Fritz Peter (1899–1949) and follows from the discussion above.

H. Weyl (1885–1955)

Theorem 6.16 (Peter–Weyl). *Let G be a compact Lie group. Then $\{\xi^j_{V,i}\}_{\xi\in\mathrm{Irr}_{\mathbb{C}}(G)}$ is a complete orthogonal basis for $L^2(G)$. Furthermore, $L^2(G)=\oplus_{\xi\in\mathrm{Irr}_{\mathbb{C}}(G)}\dim\{\xi\}\cdot\xi$ as a representation space for G under left multiplication.*

Remark 6.17 Let G be a compact Lie group which is equipped with a bi-invariant metric for G. Let E_λ be the eigenspaces of the scalar Laplacian. We use Theorem 5.11 to decompose decompose $L^2(G)$ as an orthogonal direct sum in the form $L^2(G)=\oplus_\lambda E_\lambda$. Each E_λ is a finite-dimensional representation space for G and, consequently, decomposes as the direct sum of irreducibles. Therefore, we may assume that the functions ξ^b_a are all eigenfunctions for the Laplacian. We may decompose a smooth function $\phi\in C^\infty(G)$ into a generalized Fourier series:

$$\phi=\sum_{a,b}\|\xi^b_a\|^{-2}_{L^2}(\phi,\xi^b_a)_{L^2}\xi^b_a\,.$$

One must renormalize the Fourier coefficients as the functions ξ^b_a are an orthogonal but not an orthonormal basis for L^2. This series converges in the C^∞ topology.

Theorem 6.18 *If G is a compact Lie group, then there exists a smooth representation σ which embeds G into the orthogonal group $\mathrm{SO}(p)$ for some p. Consequently, every compact Lie group can be regarded as a closed subgroup of a matrix group.*

This can fail if G is non-compact. By Lemma 6.11, $\mathrm{SL}(2,\mathbb{R})$ is homotopy equivalent to $\mathrm{SO}(2)=S^1$ and, therefore, $\pi_1(\mathrm{SL}(2,\mathbb{R}))=\mathbb{Z}$. The universal cover $\overline{\mathrm{SL}(2,\mathbb{R})}$ is an example of a group that is not a matrix group; we will show in Lemma 6.25 that this group admits no faithful finite-dimensional representation. However, locally G can always be identified with a matrix group by Theorem 6.6.

Proof. Let $\{\xi_1,\dots\}$ be an enumeration of $\mathrm{Irr}_{\mathbb{C}}(G)$; the collection of isomorphism classes of irreducible representations of G is countable by Theorem 6.16. Let $n_i:=\dim\{\xi_i\}$. Each ξ_i is a group homomorphism from G to $\mathrm{GL}_{\mathbb{C}}(n_i)$ and $\sigma_k:=\xi_1\oplus\dots\oplus\xi_k$ is a group homomorphism from G to $\mathrm{GL}_{\mathbb{C}}(n_1)\times\dots\times\mathrm{GL}_{\mathbb{C}}(n_k)\subset\mathrm{GL}_{\mathbb{R}}(m_k)$ where $m_k:=2(n_1+\dots+n_k)$. We will show that σ_k is an embedding for k large. Let $\xi_{i,a}^b$ be the complex-valued coefficient functions of each ξ_i where $1 \le a, b \le n_i$. It suffices to show that the collection of all the functions $\{\Re(\xi_{i,a}^b), \Im(\xi_{i,a}^b)\}_{1 \le i \le k}$ defines an embedding of G into a Euclidean space of the appropriate dimension; this will identify G with a closed subgroup of the matrix group $\mathrm{GL}_{\mathbb{R}}(m_k)$ and justify our restricting to matrix groups when studying compact Lie groups.

By the Whitney Embedding Theorem (see Assertion 2 of Theorem 2.2 in Book I), we may regard G as a subset of $\mathbb{R}^m$ for some m. Let $\{x^1, \dots, x^m\}$ be the coordinate functions on $\mathbb{R}^m$. We regard the $\{x^\ell\}$ as functions from G into $\mathbb{R}$. By Remark 6.5.5, we may expand each x^ℓ in a Fourier series $x^\ell = \sum_{i=1}^\infty \sum_{a,b=1}^{\dim(\xi_i)} c_b^{\ell,i,a} \xi_{i,a}^b$ which converges in the C^k topology for any k. Truncate this series to define functions $x_\mu^\ell = \sum_{i=1}^\mu \sum_{a,b=1}^{\dim(\xi_i)} c_b^{\ell,i,a} \xi_{i,a}^b$ that also define a proper embedding $\{x_\mu^1, \dots, x_\mu^m\}$ of G into $\mathbb{C}^m$ for some $\mu < \infty$. It now follows that, as desired, the functions $\{\xi_{i,a}^b\}$ for $1 \le i \le \mu$, $1 \le a, b \le \dim\{\xi_i\}$ define a proper embedding of G into $\mathbb{C}^m \subset \mathbb{R}^{2m}$. This lets us regard G as a closed subgroup of $\mathrm{GL}(p)$ for some p. By Lemma 6.12, we can choose the inner product on $\mathbb{R}^p$ to be preserved by G so $G \subset \mathrm{O}(p)$. We may double the dimension to ensure $G \subset \mathrm{SO}(2p)$. □

6.6 BI-INVARIANT PSEUDO-RIEMANNIAN METRICS

By Lemma 6.12, any compact Lie group admits a bi-invariant Riemannian metric. We now investigate bi-invariant pseudo-Riemannian metrics on non-compact Lie groups. Let $\mathrm{ad}(A)$ be the linear map which sends B to $[A, B]$. We begin with a useful observation. We shall assume G is a matrix group for the sake of simplicity; this is an inessential restriction.

Theorem 6.19 *Let G be a closed subgroup of $\mathrm{GL}(m, \mathbb{R})$ for some m.*

1. *Let $g \in S^2(T^*G)$ be a left-invariant symmetric bilinear form on TG. The following conditions are equivalent:*
   - *(a) g is bi-invariant.*
   - *(b) $g(hBh^{-1}, hCh^{-1}) = g(B, C)$ for all $h \in G$ and $B, C \in \mathfrak{g}$.*
   - *(c) ad is skew-adjoint, i.e., $g([A, B], C) + g(B, [A, C]) = 0$ for any $A, B, C \in \mathfrak{g}$.*
2. *If g is a bi-invariant pseudo-Riemannian metric, then*
   - *(a) $\nabla_A B = \frac{1}{2}[A, B]$ for $A, B \in \mathfrak{g}$.*
   - *(b) The curves $h_1 \exp^{\mathfrak{g}}(tA) h_2$ are geodesics in (G, g) for any $h_i \in G$ and $A \in \mathfrak{g}$.*
   - *(c) $R(A, B)C = -\frac{1}{4}[[A, B], C]$ for all $A, B, C \in \mathfrak{g}$.*

Proof. Since left and right multiplication are linear, $L_h' = L_h$ and $R_h' = R_h$. Assume g is bi-invariant. Let $\gamma(t) = e^{tA}$. Then

$$\begin{aligned}
&g(B,C) = g(L_{\gamma(t)}R_{\gamma(t)^{-1}}B, L_{\gamma(t)}R_{\gamma(t)^{-1}}C) = g(e^{tA}Be^{-tA}, e^{tA}Ce^{-tA}),\\
&0 = \partial_t g(B,C)|_{t=0} = g(AB - BA, C) + g(B, AC - CA)\\
&\quad = g([A,B],C) + g(B,[A,C]).
\end{aligned}$$

Suppose that $0 = g([A,B],C) + g(B,[A,C])$ for all A, B, and C. If $\xi \in T_{\mathrm{Id}}(G)$, then let $X_\xi^L(h) := h\xi$ (resp. $X_\xi^R(h) := \xi h$) be the left-invariant (resp. right-invariant) vector fields which agree with ξ at the origin. Then:

$$\begin{aligned}
&\{R_h(X_\xi^L)\}(e) = X_\xi^L(h^{-1}) \cdot h = h^{-1} \cdot \xi \cdot h \quad \text{so} \quad R_h(X_\xi^L) = X_{h^{-1}\xi h}^L,\\
&\{L_h(X_\xi^R)\}(e) = h \cdot X_\xi^R(h^{-1}) = h \cdot \xi \cdot h^{-1} \quad \text{so} \quad L_h(X_\xi^R) = X_{h\xi h^{-1}}^R .
\end{aligned}$$

Consequently, g is bi-invariant if and only if $g(hBh^{-1}, hAh^{-1}) = g(B,A)$ for all $A, B \in \mathfrak{g}$ and all $h \in G$. Since G is connected, elements of the form e^{tC} generate G for C in $\mathfrak{g}$. Set

$$f(t,A,B,C) := g(e^{tC}Be^{-tC}, e^{tC}Ae^{-tC}), \quad A_t := e^{tC}Ae^{-tC}, \quad B_t = e^{tC}Be^{-tC}.$$

Then $f'(t,A,B,C) = g([C,B_t],A_t) + g(B_t,[C,A_t]) = 0$. This establishes Assertion 1.

Let ∇ be the Levi–Civita connection of a bi-invariant pseudo-Riemannian metric g. If $A, B, C \in \mathfrak{g}$, then the Koszul formula of Theorem 3.7 in Book I simplifies to become:

$$g(\nabla_A B, C) = \tfrac{1}{2}\{g([A,B],C) - g(A,[B,C]) + g([C,A],B)\}.$$

As g is bi-invariant, $g([C,A],B) + g(A,[C,B]) = 0$ by Assertion 1. Thus, the Koszul formula simplifies to yield $g(\nabla_A B, C) = \frac{1}{2}g([A,B],C)$ which proves Assertion 2-a. Setting $A = B$ yields $\nabla_A A = 0$. Let $\gamma(t) = \exp^{\mathfrak{g}}(t\xi)$. Then $\dot\gamma = X_\xi^L$ so $\ddot\gamma = \nabla_{X_\xi^L} X_\xi^L$ and the curves $\gamma(t)$ are geodesics. Assertion 2-b now follows. We use the Jacobi identity to complete the proof:

$$\begin{aligned}
R(A,B)C &= \nabla_A\nabla_B C - \nabla_B\nabla_A C - \nabla_{[A,B]}C\\
&= \tfrac{1}{4}[A,[B,C]] - \tfrac{1}{4}[B,[A,C]] - \tfrac{1}{2}[[A,B],C]\\
&= \tfrac{1}{4}[A,[B,C]] + \tfrac{1}{4}[B,[C,A]] - \tfrac{1}{2}[[A,B],C] = -\tfrac{1}{4}[C,[A,B]] - \tfrac{1}{2}[[A,B],C]\\
&= \tfrac{1}{4}[[A,B],C] - \tfrac{1}{2}[[A,B],C] = -\tfrac{1}{4}[[A,B],C].
\end{aligned}$$

□

We use Theorem 6.19 to establish the following result; Lemma 6.25 shows that this result can fail if G is not compact as the exponential map for $\mathrm{SL}(2,\mathbb{R})$ is not surjective.

Corollary 6.20 If G is a compact connected Lie group, then $\exp^{\mathfrak{g}} : T_eG \to G$ is surjective.

Proof. Let G be a compact connected Lie group. By Theorem 6.18, G can be regarded as a subgroup of $\mathrm{SO}(p)$ for some p. By Lemma 6.12, G admits a bi-invariant Riemannian metric g. Consequently, Theorem 6.19 shows that $\exp^{\mathfrak{g}}$ can be identified with the Riemannian exponential map defined by g. Since (G,g) is a compact metric space, it is a complete metric space. Since G is connected, the Hopf–Rinow Theorem (see Theorem 3.15 in Book I) shows that the exponential map is surjective. □

6.7 THE KILLING FORM

The following concepts were first introduced by the German mathematician Wilhelm Karl Joseph Killing.

W. Killing (1847–1923)

If G is a Lie group, let $\mathfrak{g}$ be the Lie algebra of left-invariant vector fields on G. If $\xi \in \mathfrak{g}$, let $\operatorname{ad}(\xi) : \eta \to [\xi, \eta]$; $\operatorname{ad}(\xi) \in \operatorname{Hom}(\mathfrak{g})$. One defines the *Killing form* (see Cartan [7]) setting:

$$K(\xi_1, \xi_2) := \operatorname{Tr}\{\operatorname{ad}(\xi_1)\operatorname{ad}(\xi_2)\}\,.$$

The Killing form is a symmetric bilinear form on $\mathfrak{g}$ which is invariant under left multiplication in the group. Since left and right multiplication commute, R_g is a Lie algebra isomorphism of $\mathfrak{g}$ and preserves K as well. Therefore, K is a bi-invariant symmetric bilinear form on $\mathfrak{g}$. Let $\{e_i\}$ be a basis for $\mathfrak{g}$. We can expand $[e_i, e_j] = c_{ij}{}^k e_k$ to define the *Lie algebra structure constants* of Equation (6.2.b). We then have $K(e_i, e_j) = c_{i\ell}{}^k c_{jk}{}^\ell$.

Of particular interest is the case when G is a matrix group. Let $\mathfrak{g}$ be the Lie algebra of left-invariant vector fields. If $\xi, \eta \in \mathfrak{g}$, the map $(\xi, \eta) \to \operatorname{Tr}\{\xi\eta\}$ defines a left-invariant bilinear form. We apply Theorem 6.19 to see that this form is bi-invariant by computing:

$$\operatorname{Tr}\{g^{-1}\xi g \cdot g^{-1}\eta\} = \operatorname{Tr}\{g^{-1}\xi\eta g\} = \operatorname{Tr}\{\xi\eta g g^{-1}\} = \operatorname{Tr}\{\xi\eta\}\,.$$

Let $\{e^i\}$ be an orthonormal basis for $\mathbb{R}^m$. If $\xi \in \mathfrak{so}(m)$, then

$$\operatorname{Tr}\{\xi^2\} = \sum_i (\xi^2 e^i, e^i) = -\sum_i (\xi e^i, \xi e^i)\,.$$

Consequently, $\operatorname{Tr}\{\xi^2\} \leq 0$ and $\operatorname{Tr}\{\xi^2\} = 0$ if and only if $\xi = 0$. Consequently, this bilinear form is a bi-invariant negative pseudo-Riemannian metric on $\mathrm{SO}(m)$.

Theorem 6.21

1. *If $\mathfrak{g}$ is a Lie subalgebra of $\mathfrak{so}(m)$ for some m, then K is negative semi-definite.*
2. *If G is a compact connected Lie group, then K is negative semi-definite.*
3. *If K is negative definite and if G is connected, then G is compact.*

Proof. Let $\mathfrak{g}$ be a Lie subalgebra of $\mathfrak{so}(m)$. Let $\langle \xi, \eta \rangle = -\operatorname{Tr}\{\xi, \eta\}$ be a positive definite inner product on $\mathfrak{so}(m)$ and hence on $\mathfrak{g}$. We compute

$$\begin{aligned}\langle \operatorname{ad}(\xi)\eta_1, \eta_2 \rangle &= -\operatorname{Tr}\{(\xi\eta_1 - \eta_1\xi)\eta_2\} = -\operatorname{Tr}\{\xi\eta_1\eta_2 - \eta_1\xi\eta_2\} = -\operatorname{Tr}\{\eta_1\eta_2\xi - \eta_1\xi\eta_2\} \\ &= \operatorname{Tr}\{\eta_1[\xi, \eta_2]\} = -\langle \eta_1, \operatorname{ad}(\xi)\eta_2 \rangle\,.\end{aligned}$$

This implies that $\operatorname{ad}(\xi)$ is skew-adjoint with respect to the positive definite inner product $\langle \cdot, \cdot \rangle$. It follows that the eigenvalues of $\operatorname{ad}(\xi)$ are all purely imaginary and, therefore, $\operatorname{Tr}\{\operatorname{ad}(\xi)^2\}$ is non-positive. This establishes Assertion 1.

Suppose G is a compact connected Lie group. By Theorem 6.18, there exists a smooth representation σ embedding G into $\operatorname{GL}_{\mathbb{R}}(m)$ for some m. By Lemma 6.12, we may assume G preserves some positive definite inner product on $\mathbb{R}^m$; since G is connected, we may assume that G is a closed subgroup of $\operatorname{SO}(m)$ and, consequently, the Lie algebra $\mathfrak{g}$ is a Lie subalgebra of $\mathfrak{so}(m)$. Assertion 2 now follows from Assertion 1.

We shall show presently in Lemma 7.13 that G is complete. Since the Killing form is negative definite, $\operatorname{ad}(A)$ is non-zero for every $A \in G$. This implies that the adjoint action

$$\operatorname{Ad}(g) : A \to g^{-1}Ag$$

gives a map from G to $\operatorname{Hom}(\mathfrak{g})$ which is an embedding of a neighborhood of Id in G into $\operatorname{GL}(\mathfrak{g})$; the analysis of Theorem 6.19 was purely local so the conclusions concerning the exponential map continue to hold. Let $g = -K$; this is bi-invariant and positive definite. Let ρ be the associated Ricci tensor. Then:

$$\begin{aligned}\rho(A, A) &:= \operatorname{Tr}\{C \to R(C, A)A\} = -\operatorname{Tr}\{C \to \tfrac{1}{4}[[C, A], A]\} \\ &= -\tfrac{1}{4}\operatorname{Tr}\{C \to [A, [A, C]]\} = -\tfrac{1}{4}\operatorname{Tr}\{C \to \operatorname{ad}(A)\operatorname{ad}(A)C\} \\ &= -\tfrac{1}{4}\operatorname{Tr}\{\operatorname{ad}(A)^2\} = -\tfrac{1}{4}K(A, A)\,.\end{aligned} \tag{6.7.a}$$

Equation (6.7.a) shows the Ricci tensor is uniformly positive. We have already shown that G is complete. By assumption, G is connected. It now follows that G is compact by Myers' Theorem (see Theorem 3.20 in Book I). □

6.7.1 THE KILLING FORM OF THE ORTHOGONAL GROUP. Let $(V, \langle \cdot, \cdot \rangle)$ be an inner product space of signature (p, q) and let $G = \operatorname{O}(V, \langle \cdot, \cdot \rangle)$ be the associated orthogonal group. We suppose $p > 0$ and $q > 0$ so G has four connected components.

Lemma 6.22 Let $G = \operatorname{O}(V, \langle \cdot, \cdot \rangle)$ where $\langle \cdot, \cdot \rangle$ has signature (p, q). Let $m = \dim\{V\} = p + q$.

1. $K(\xi, \eta) = (m - 2)\operatorname{Tr}\{\xi\eta\}$ has signature $(p(p-1) + q(q-1), pq)$.
2. If $m \neq 4$ and if Θ is a bi-invariant symmetric bilinear form on $\mathfrak{g}$, then $\Theta = \lambda K$ for some λ. Therefore, in particular, G does not admit a bi-invariant Riemannian metric if $p > 0$, $q > 0$, $(p, q) \neq (1, 1)$, and $m \neq 4$.

Proof. Let $\mathfrak{g} = \{A \in \text{Hom}(V) : \langle Av, w\rangle + \langle v, Aw\rangle = 0\}$ be the Lie algebra of $\text{O}(V, (\cdot,\cdot))$. Let $\{e_1, \dots, e_m\}$ be an orthonormal basis for V. Consequently, $\langle e_i, e_j\rangle = 0$ for $i \neq j$ while there is a choice of signs $\epsilon_i = \pm 1$ so $\langle e_i, e_i\rangle = \epsilon_i$. Set

$$A_{ij}e_k = \left\{ \begin{array}{ll} e_i & \text{if } k = j \\ -\epsilon_i\epsilon_j e_j & \text{if } k = i \\ 0 & \text{otherwise} \end{array} \right\}.$$

If $\{i, j, k, \ell\}$ are distinct indices, then $A_{ij}A_{k\ell} = 0$ so $\text{Tr}\{A_{ij}A_{k\ell}\} = 0$. If $\{i, j, k\}$ are distinct indices, then

$$A_{ij}A_{ik}e_k = A_{ij}e_i = -\epsilon_i\epsilon_j e_j, \quad A_{ij}A_{ik}e_i = -\epsilon_i\epsilon_k A_{ij}e_k = 0, \quad \text{Tr}\{A_{ij}A_{ik}\} = 0\,.$$

Finally, $A_{ij}A_{ij}e_j = A_{ij}e_i = -\epsilon_i\epsilon_j e_j$ and $A_{ij}A_{ij}e_i = -\epsilon_i\epsilon_j A_{ij}e_j = -\epsilon_i\epsilon_j e_i$ so

$$\text{Tr}\{A_{ij}A_{ij}\} = -2\epsilon_i\epsilon_j\,.$$

This shows that $\text{Tr}\{\xi\eta\}$ is a non-degenerate bi-invariant pseudo-Riemannian metric on G of signature $(p(p-1) + q(q-1), pq)$. Let $\{i, j, k\}$ be distinct indices. The only non-trivial commutators are of the form $[A_{ij}, A_{ik}]$. We compute:

$$\begin{aligned} (A_{ij}A_{ik} - A_{ik}A_{ij})e_k &= -\epsilon_i\epsilon_j e_j = -\epsilon_i\epsilon_j A_{jk}e_k, \\ (A_{ij}A_{ik} - A_{ik}A_{ij})e_j &= \epsilon_i\epsilon_k e_k = -\epsilon_i\epsilon_j A_{jk}e_j\,, \end{aligned}$$

so $[A_{ij}, A_{ik}] = -\epsilon_i\epsilon_j A_{jk}$. If $\{i, j, k, \ell\}$ are distinct indices, $K(A_{ij}, A_{k\ell}) = 0$. We examine:

$$\begin{aligned} \text{ad}(A_{ij})\,\text{ad}(A_{ik})A_{ij} &= \star\,\text{ad}(A_{ij})A_{kj} = \star A_{ik}, \\ \text{ad}(A_{ij})\,\text{ad}(A_{ik})A_{in} &= \star\,\text{ad}(A_{ij})A_{kn} = 0, \\ \text{ad}(A_{ij})\,\text{ad}(A_{ik})A_{jk} &= \star\,\text{ad}(A_{ij})A_{ij} = 0, \\ \text{ad}(A_{ij})\,\text{ad}(A_{ik})A_{nk} &= \star\,\text{ad}(A_{ij})A_{in} = \star A_{jn}\,, \end{aligned}$$

where $\star$ is a $\pm$ sign not of interest. This shows $K(A_{ij}, A_{ik}) = 0$. Finally, we compute

$$\begin{aligned} \text{ad}(A_{ij})\,\text{ad}(A_{ij})A_{ik} &= -\epsilon_i\epsilon_j\,\text{ad}(A_{ij})A_{jk} = \text{ad}(A_{ji})A_{jk} = -\epsilon_j\epsilon_i A_{ik}, \\ \text{ad}(A_{ij})\,\text{ad}(A_{ij})A_{jk} &= -\epsilon_i\epsilon_j\,\text{ad}(A_{ij})\,\text{ad}(A_{ji})A_{jk} = \text{ad}(A_{ij})A_{ik} = -\epsilon_i\epsilon_j A_{jk}\,. \end{aligned}$$

Consequently, $K(\xi, \eta) = -2(m-2)\epsilon_i\epsilon_j = (m-2)\,\text{Tr}\{\xi\eta\}$ and Assertion 1 follows.

Let Θ be a bi-invariant symmetric bilinear form on $\mathfrak{g}$. If $m = 2$, then $\mathfrak{g} = \mathbb{R}$ and Θ is unique. Suppose $m \geq 3$. Let $\{i, j, k\}$ be distinct indices. Suppose $\epsilon_j = \epsilon_k$. We define the rotation $T^{\theta}_{j,k}$ in G by setting:

$$T_{j,k,\theta}e_\ell = \left\{ \begin{array}{ll} \cos\theta e_j + \sin\theta e_k & \text{if } \ell = j \\ -\sin\theta e_j + \cos\theta e_k & \text{if } \ell = k \\ e_\ell & \text{otherwise} \end{array} \right\}.$$

We have $T_{j,k,\theta}^{-1} A_{ij} T_{j,k,\theta} = \cos\theta A_{ij} + \sin\theta A_{ik}$. Therefore,

$$\begin{aligned}\Theta(A_{ij}, A_{ij}) &= \Theta(T_{j,k,\theta}^{-1} A_{ij} T_{j,k,\theta}, T_{j,k,\theta}^{-1} A_{ij} T_{j,k,\theta}) \\ &= \cos^2\theta\, \Theta(A_{ij}, A_{ij}) + 2\cos\theta\sin\theta\, \Theta(A_{ij}, A_{ik}) + \sin^2\theta\, \Theta(A_{ik}, A_{ik}).\end{aligned}$$

This equality for all θ implies $\Theta(A_{ij}, A_{ij}) = (A_{ik}, A_{ik})$ and $\Theta(A_{ij}, A_{ik}) = 0$. If $\epsilon_j \neq \epsilon_k$ we replace cos and sin by cosh and sinh and change a sign to define the *hyperbolic boost*

$$T_{j,k,\theta} e_\ell = \left\{ \begin{array}{ll} \cosh\theta e_j + \sinh\theta e_k & \text{if } \ell = j \\ \sinh\theta e_j + \cosh\theta e_k & \text{if } \ell = k \\ e_\ell & \text{otherwise} \end{array} \right\}.$$

The same argument shows

$$\Theta(A_{ij}, A_{ij}) = \cosh^2\theta\, \Theta(A_{ij}, A_{ij}) + 2\cosh\theta\sinh\theta\, \Theta(A_{ij}, A_{ik}) + \sinh^2\theta\, \Theta(A_{ik}, A_{ik}).$$

We now conclude $\Theta(A_{ij}, A_{ij}) = -\Theta(A_{ik}, A_{ik})$ and $\Theta(A_{ij}, A_{ik}) = 0$. Combining these two calculations yields $\epsilon_i\epsilon_j\Theta(A_{ij}, A_{ij}) = \epsilon_i\epsilon_k\Theta(A_{ik}, A_{ik})$ and $\Theta(A_{ij}, A_{ik}) = 0$. And permuting the indices recursively lets us show:

$$\epsilon_i\epsilon_j\Theta(A_{ij}, A_{ij}) = \epsilon_u\epsilon_v\Theta(A_{uv}, A_{uv}) \quad \text{and} \quad \Theta(A_{ij}, A_{ik}) = 0 \quad \text{for} \quad \{i, j, k\} \quad \text{distinct}.$$

This shows Θ is unique if $m = 3$. For $m > 4$, it only remains to examine $\Theta(A_{ij}, A_{k\ell})$ for $\{i, j, k, \ell\}$ distinct. Let $\{i, j, k, \ell, n\}$ be distinct indices. We suppose $\epsilon_\ell = \epsilon_n$ as the other case is similar. We compute

$$\Theta(A_{ij}, A_{k\ell}) = \Theta(T_{\ell,n,\theta}^{-1} A_{ij} T_{\ell,n,\theta}, T_{\ell,n,\theta}^{-1} A_{k\ell} T_{\ell,n,\theta}) = \cos\theta\, \Theta(A_{ij}, A_{k\ell}) + \sin\theta\, \Theta(A_{ij}, A_{k\ell}).$$

This identity for all θ implies $\Theta(A_{ij}, A_{k\ell}) = 0$ and completes the determination of Θ. □

We shall show presently in Lemma 6.24 that if $m = 4$, then the universal cover of SO(4) is $S^3 \times S^3$ and, consequently, $\mathfrak{so}(4) = \mathfrak{so}(3) \oplus \mathfrak{so}(3)$ and the space of bi-invariant symmetric bilinear forms has dimension 2; 4-dimensional geometry is very different in this regard.

6.7.2 THE KILLING FORM OF THE $AX + B$ GROUP. The following is a useful example. We consider the set of linear transformations $f_{u,v}(x) = ux + v$, for $u > 0$, which we identify with $\mathbb{R}^+ \times \mathbb{R}$. Then:

$$(f_{u,v} \circ f_{\tilde{u},\tilde{v}})(x) = u\tilde{u}x + (u\tilde{v} + v) = f_{u\tilde{u}, u\tilde{v}+v}.$$

We give $\mathbb{R}^* \times \mathbb{R}$ a group structure by setting $(u, v) \star (\tilde{u}, \tilde{v}) = (u\tilde{u}, u\tilde{v} + v)$.

Lemma 6.23 Let G be the $ax+b$ group.

1. There is a basis for the Lie algebra $\mathfrak{g}$ so
$$[X^L,Y^L]=Y^L,\ K(X^L,X^L)=1,\ K(X^L,Y^L)=0,\text{ and } K(Y^L,Y^L)=0\,.$$
This group does not admit a bi-invariant measure.

2. If $\sigma:G\to \mathrm{SO}(m)$ is a group homomorphism, then $\sigma_*:\mathfrak{g}\to\mathfrak{so}(m)$ is not injective. In particular, the canonical representation of G on $\mathbb{R}^2$ does not admit an invariant positive definite inner product.

Proof. We have:

$$\begin{array}{rrr}
L_{(a,b)}(u,v)=(au,av+b), & (L_{a,b})_*(\partial_u)=a\partial_u, & L_{a,b}^*(du)=adu,\\
 & (L_{a,b})_*(\partial_v)=a\partial_v, & L_{a,b}^*(dv)=adv,\\
R_{(a,b)}(u,v)=(au,v+ub), & (R_{a,b})_*(\partial_u)=a\partial_u+b\partial_v, & R_{a,b}^*(du)=adu,\\
 & (R_{a,b})_*(\partial_v)=\partial_v, & R_{a,b}^*(dv)=dv+bdu\,.
\end{array}$$

Let $X^{L/R}$ (resp. $Y^{L/R}$) be the left/right-invariant vector field agreeing with ∂_u (resp. ∂_v) at $(1,0)$, and let $\omega_X^{L/R}$ (resp. $\omega_Y^{L/R}$) be the left/right-invariant 1-form agreeing with du (resp. dv) at $(1,0)$, then

$$\begin{array}{rrrr}
X^L(u,v)=u\partial_u, & Y^L(u,v)=u\partial_v, & \omega_X^L=u^{-1}du, & \omega_Y^L=u^{-1}dv,\\
{}[X^L,Y^L]=Y^L, & K(X^L,X^L)=1, & K(X^L,Y^L)=0, & K(Y^L,Y^L)=0,\\
d\omega_X^L=0, & d\omega_Y^L=-\omega_X^L\wedge\omega_Y^L, & & \\
X^R(u,v)=u\partial_u+v\partial_v, & Y^R(u,v)=\partial_v, & \omega_X^R=u^{-1}du, & \omega_Y^R=dv-u^{-1}du,\\
{}[X^R,Y^R]=-Y^R, & K(X^R,X^R)=1, & K(X^R,Y^R)=0, & K(Y^R,Y^R)=0,\\
d\omega_X^R=0, & d\omega_Y^R=\omega_X^R\wedge\omega_Y^R\,. & &
\end{array}$$

The left-invariant volume element can be taken to be given by $\omega_X^L\wedge\omega_Y^L=u^{-2}dudv$. The right-invariant volume element can be taken to be $\omega_X^R\wedge\omega_Y^R=u^{-1}dudv$. These are different; there is no bi-invariant volume form and hence no bi-invariant pseudo-Riemannian metric. The action of $R_{a,b}^*$ on $\Lambda^1(\mathfrak{g})$ is given by:

$$\begin{aligned}
R_{a,b}^*(\omega_X^L) &= R_{a,b}^*(u^{-1}du)=\omega_X^L,\\
R_{a,b}^*(\omega_Y^L) &= R_{a,b}^*(u^{-1}dv)=(au)^{-1}(dv+bdu)\\
&= ba^{-1}u^{-1}du+au^{-1}dv=ba^{-1}\omega_X^L+a^{-1}\omega_Y^L,\\
R_{a,b}^*(\omega_X^L\wedge\omega_Y^L) &= a^{-1}\omega_X^L\wedge\omega_X^L\,.
\end{aligned}$$

The action of $R^*_{a,b}$ on $\Lambda^1(\mathfrak{g})$ defines a group homomorphism from G to $M_2(\mathbb{R})$ and exhibits G as a matrix group:

$$\sigma(a,b) := \begin{pmatrix} 1 & ba^{-1} \\ 0 & a^{-1} \end{pmatrix},$$

$$\sigma((a,b) * (c,d)) = \sigma(ac, b+ad) = \begin{pmatrix} 1 & a^{-1}c^{-1}b + c^{-1}d \\ 0 & a^{-1}c^{-1} \end{pmatrix},$$

$$\sigma(a,b)\sigma(c,d) = \begin{pmatrix} 1 & ba^{-1} \\ 0 & a^{-1} \end{pmatrix} \begin{pmatrix} 1 & dc^{-1} \\ 0 & c^{-1} \end{pmatrix}.$$

We will return to the $ax+b$ group in Section 7.4.3 when we construct a left-invariant metric on this group which is Lorentzian and incomplete. We conclude by remarking that

$$K(X,X) = 1, \quad K(X,Y) = 0, \quad K(Y,Y) = 0$$

so the Killing form is positive semi-definite and degenerate in this instance. This proves Assertion 1. Theorem 6.21 implies $\mathfrak{g}$ is not a Lie subalgebra of $\mathfrak{so}(m)$ for any m so there is no σ so σ_* is injective. Assertion 2 follows. □

6.8 THE CLASSICAL GROUPS IN LOW DIMENSIONS

There are some relationships among the low-dimensional Lie groups which are important.

Lemma 6.24

1. $\mathrm{SO}(3) = S^3/\mathbb{Z}_2 = \mathbb{RP}^3$ where $\mathbb{Z}_2 = \{\pm 1\}$. The round metric on S^3 induces a bi-invariant Riemannian metric on SO(3).
2. $\mathrm{SU}(2) = S^3$. The round metric on S^3 induces a bi-invariant Riemannian metric on SU(2).
3. $\mathrm{SO}(4) = (S^3 \times S^3)/\mathbb{Z}_2$ where $\mathbb{Z}_2 = \{\pm 1\}$ acts diagonally on $S^3 \times S^3$. There is a 2-parameter family of bi-invariant Riemannian metrics on SO(4) corresponding to rescaling the round metrics on each factor of $S^3 \times S^3$.

Proof. Let $\mathbb{H}$ be the quaternions. If $g \in S^3$, let $\tau(g)x = gx\bar{g}$. Since

$$\|\tau(g)\|^2 = gx\bar{g}\overline{gx\bar{g}} = gx\bar{g}g\bar{x}\bar{g} = gx\bar{x}\bar{g} = g\|x\|^2\bar{g} = \|x\|^2 g\bar{g} = \|x\|^2,$$

$\tau(g) \in \mathrm{O}(4)$. Since $\tau(g)1 = 1$, $\tau(g)$ restricts to define a map $\sigma(g) : 1^\perp \to 1^\perp$. Identify

$$\mathbb{R}^3 = x^1 i + x^2 j + x^3 k$$

with the purely imaginary quaternions. We then have $\sigma : S^3 \to \mathrm{O}(4)$. Since S^3 is connected, σ is a map from S^3 to $\mathrm{SO}(3)$. Let

$$g_i(\theta) = \cos\theta + \sin\theta i, \quad g_j(\theta) = \cos\theta + \sin\theta j, \quad g_k(\theta) = \cos\theta + \sin\theta k\,.$$

Let $\{A_{12}, A_{23}, A_{31}\}$ be the basis for $\mathfrak{so}(3)$ defined above and let $\{A_i = i, A_j = j, A_k = k\}$ be the natural basis for the Lie algebra of S^3. Then

$$\begin{aligned}
\sigma(g_i(\theta))i &= (\cos\theta + \sin\theta i)i(\cos\theta - \sin\theta i) = i,\\
\sigma(g_i(\theta))j &= (\cos\theta + \sin\theta i)j(\cos\theta - \sin\theta i) = \cos(2\theta)j + \sin(2\theta)k,\\
\sigma(g_i(\theta))k &= (\cos\theta + \sin\theta i)k(\cos\theta - \sin\theta i) = -\sin(2\theta)j + \cos(2\theta)k,\\
\sigma(g_j(\theta))i &= (\cos\theta + \sin\theta j)i(\cos\theta - \sin\theta j) = \cos(2\theta)i - \sin(2\theta)k,\\
\sigma(g_j(\theta))j &= (\cos\theta + \sin\theta j)j(\cos\theta - \sin\theta j) = j,\\
\sigma(g_j(\theta))k &= (\cos\theta + \sin\theta j)k(\cos\theta - \sin\theta j) = \sin(2\theta)i + \cos(2\theta)k,\\
\sigma(g_k(\theta))i &= (\cos\theta + \sin\theta k)i(\cos\theta - \sin\theta k) = \cos(2\theta)i + \sin(2\theta)j,\\
\sigma(g_k(\theta))j &= (\cos\theta + \sin\theta k)j(\cos\theta - \sin\theta k) = -\sin(2\theta)i + \cos(2\theta)j,\\
\sigma(g_k(\theta))k &= (\cos\theta + \sin\theta k)k(\cos\theta - \sin\theta k) = k,\\
\sigma_*(A_i) &= \dot\sigma(g_i)(0) = -2A_{23},\ \sigma_*(A_j) = \dot\sigma(g_i)(0) = -2A_{31}, \sigma_*(A_k) = -2A_{12}\,.
\end{aligned}$$

This shows that σ_* is a Lie group isomorphism from the Lie algebra of S^3 to $\mathfrak{so}(3)$ and hence σ is a diffeomorphism from a neighborhood of $1 \in S^3$ to a neighborhood of Id in SO(3). Consequently, σ is an open map on a neighborhood of 1. Since

$$\sigma(gh) = \sigma(g)\sigma(h)\,,$$

σ is an open map. Since S^3 is compact, σ is a closed map. Since SO(3) is connected, σ is surjective. Therefore, σ is a local diffeomorphism from S^3 onto SO(3) and hence is a covering projection. Suppose $\sigma(x) = \mathrm{Id}$. Then $xy\bar{x} = y$ for all $y \in S^3$ and hence $xy = yx$ for all $y \in S^3$. This implies x is scalar so $x = \pm 1$ and $\ker\{\sigma\} = \pm 1$. Consequently,

$$\mathrm{SO}(3) = S^3/\mathbb{Z}_2$$

is the sphere with antipodal points identified. This implies that $\mathrm{SO}(3) = \mathbb{RP}^3$ so S^3 has a unique bi-invariant metric which is easily seen to be the round metric.

We now study SU(2). Let

$$e_0 = \begin{pmatrix} 1 & 0\\ 0 & 1\end{pmatrix},\ e_1 = \begin{pmatrix} 0 & 1\\ -1 & 0\end{pmatrix},\ e_2 = \begin{pmatrix} 0 & i\\ i & 0\end{pmatrix},\ e_3 = \begin{pmatrix} i & 0\\ 0 & -i\end{pmatrix}. \tag{6.8.a}$$

Let $\alpha := x^0 + \sqrt{-1}e^3$ and $\beta := x^1 + \sqrt{-1}e^2$ with

$$\alpha\bar\alpha + \beta\bar\beta = (x^0)^2 + (x^1)^2 + (x^2)^2 + (x^3)^2 = 1\,.$$

We decompose $g \in \mathrm{SU}(2)$ in the form:

$$g = \begin{pmatrix} \alpha & \beta \\ -\bar{\beta} & \bar{\alpha} \end{pmatrix} = x^0 e_0 + x^1 e_1 + x^2 e_2 + x^3 e_3 .$$

We establish the required identification of SU(2) with S^3 by showing that the quaternion relations are satisfied:

$$e_1^2 = e_2^2 = e_3^2 = -\mathrm{Id},\ e_1 e_2 = -e_2 e_1 = e_3,\ e_2 e_3 = -e_3 e_2 = e_1,\ e_3 e_1 = -e_1 e_3 = e_2 . \quad (6.8.b)$$

We now examine SO(4). Let $\tau(g_1, g_2) g := g_1 g g_2$ for $g_1, g_2, g \in S^3$. We compute:

$$\|\tau(g_1, g_2) g\|^2 = g_1 g g_2 \bar{g}_2 \bar{g} \bar{g}_1 = 1$$

and, consequently, τ defines a representation from $S^3 \times S^3$ to SO(4). If $\tau(g_1, g_2) = \mathrm{Id}$, then $\tau(g_1, g_2) 1 = 1$ and, therefore, $g_1 \bar{g}_2 = 1$ so $g_1 = g_2$. The argument given above to study SO(3) then shows $g_1 = g_2 = \pm 1$ so

$$\tau : (S^3 \times S^3)/\{\pm(1,1)\} \to \mathrm{SO}(4)$$

is injective. It is smooth. Let ξ be a left-invariant vector field on $S^3 \times S^3$. If $\tau_* \xi = 0$, then the flow for $\tau_* \xi$ is trivial so $\tau(e^{t\xi}) = \mathrm{Id}$; this is not possible as τ is locally injective. This implies τ_* is injective. Since $\dim\{S^3 \times S^3\} = \dim\{\mathrm{SO}(4)\} = 6$, we conclude τ is a covering projection so $\mathrm{SO}(4) = (S^3 \times S^3)/\{\pm(1,1)\}$. □

Let $\mathrm{SL}(2,\mathbb{R})$ be the group of invertible 2×2 invertible real matrices of determinant 1. There is a close relationship between $\mathrm{SL}(2,\mathbb{R})$ and SU(2) which we will exploit; this relates the Clifford algebra on a metric of signature $(0,3)$ with the Clifford algebra on a metric of signature $(2,1)$.

Lemma 6.25

1. The Killing form is a bi-invariant Lorentzian metric on $\mathrm{SL}(2,\mathbb{R})$, $\mathrm{SL}(2,\mathbb{R})$ admits a bi-invariant volume form, and $\mathrm{SL}(2,\mathbb{R})$ is geodesically complete.

2. No non-trivial representation of $\mathrm{SL}(2,\mathbb{R})$ admits an invariant Euclidean inner product.

3. The exponential map $\exp^{\mathfrak{g}}$ is not surjective.

4. $\pi_1(\mathrm{SL}(2,\mathbb{R})) = \mathbb{Z}$.

5. Let $\overline{\mathrm{SL}(2,\mathbb{R})}$ be the universal covering group of $\mathrm{SL}(2,\mathbb{R})$. Then no finite-dimensional representation of $\overline{\mathrm{SL}(2,\mathbb{R})}$ is faithful so $\overline{\mathrm{SL}(2,\mathbb{R})}$ is not a matrix group.

Proof. Let $\{e_0, e_1, e_2, e_3\}$ be as defined in Equation (6.8.a); these satisfy the quaternion relations of Equation (6.8.b). Let $f_0 = e_0$, $f_1 = e_1$, $f_2 = -\sqrt{-1}e_2$, and $f_3 = -\sqrt{-1}e_3$. Then:

$$f_0 = \begin{pmatrix} 1 & 0 \\ 0 & 1 \end{pmatrix}, \quad f_1 = \begin{pmatrix} 0 & 1 \\ -1 & 0 \end{pmatrix}, \quad f_2 = \begin{pmatrix} 0 & 1 \\ 1 & 0 \end{pmatrix}, \quad f_3 = \begin{pmatrix} 1 & 0 \\ 0 & -1 \end{pmatrix}.$$

The f_i's satisfy the Clifford commutation relations for a Clifford algebra of signature $(2, 0)$:

$$\begin{array}{llll} f_i f_j + f_j f_i = 0 & \text{for } 1 \le i < j \le 3, & f_1^2 = -\operatorname{Id}, & f_2^2 = f_3^2 = \operatorname{Id}, \\ f_1 f_2 = f_3, & [f_1, f_2] = 2f_3, & [f_2, f_3] = -2f_1, & [f_3, f_1] = 2f_2\,. \end{array}$$

Let $e(x) = x^i e_i$ and let $f(x) = y^i f_i$. Then:

$$e(x) = \begin{pmatrix} x^0 + ix^3 & x^1 + ix^2 \\ -x^1 + ix^2 & x^0 - ix^3 \end{pmatrix} \quad \text{and} \quad f(x) = \begin{pmatrix} y^0 + y^3 & y^1 + y^2 \\ -y^1 + y^2 & y^0 - y^3 \end{pmatrix}.$$

Let $(x, \tilde{x}) := x^0\tilde{x}^0 + x^1\tilde{x}^1 + x^2\tilde{x}^2 + x^3\tilde{x}^3$ and let $\langle y, \tilde{y}\rangle := y^0\tilde{y}^0 + y^1\tilde{y}^1 - y^2\tilde{y}^2 - y^3\tilde{y}^3$. Then

$$\mathrm{SU}(2) := \{e(x) : (x, x) = 1\} \quad \text{and} \quad \mathrm{SL}(2, \mathbb{R}) := \{f(y) : \langle y, y\rangle = 1\}.$$

We extend $e(\cdot)$ to be complex linear and $(\cdot, \cdot)$ to be complex bi-linear. Let

$$\mathrm{SU}_{\mathbb{C}}(2) := \{e(z) : (z, z) = 1\} \subset \mathrm{GL}(2, \mathbb{C}).$$

Then both SU(2) and $\mathrm{SL}(2, \mathbb{R})$ are Lie subgroups of $\mathrm{SU}_{\mathbb{C}}(2)$ and we pass from SU(2) to $\mathrm{SL}(2, \mathbb{R})$ by setting $x^2 = iy^2$ and $x^3 = iy^3$. We have

$$\begin{array}{lll} \operatorname{ad}(e_1)e_1 = 0, & \operatorname{ad}(e_1)e_2 = 2e_3, & \operatorname{ad}(e_1)e_3 = -2e_2, \\ \operatorname{ad}(e_2)e_1 = -2e_3, & \operatorname{ad}(e_2)e_2 = 0, & \operatorname{ad}(e_2)e_3 = 2e_1, \\ \operatorname{ad}(e_3)e_1 = 2e_2, & \operatorname{ad}(e_3)e_2 = -2e_1, & \operatorname{ad}(e_3)e_3 = 0\,. \end{array}$$

It now follows that $K(e_i, e_j) = -8\delta_{ij}$ so the Killing form is a negative definite form on SU(2). Owing to the factor of i defining f_2 and f_3, the Killing form is a form of signature $(2, 1)$ on $\mathrm{SL}(2, \mathbb{R})$. The Killing form is a non-degenerate bi-invariant pseudo-Riemannian metric. Consequently, by Theorem 6.19, the metric exponential map and the group exponential map coincide. Therefore, all geodesics extend for infinite time and $\mathrm{SL}(2, \mathbb{R})$ is geodesically complete with respect to the metric defined by the Killing form.

By Theorem 6.21, if $\mathfrak{sl}(2, \mathbb{R})$ is a Lie subalgebra of $\mathfrak{so}(m)$ for some m, then the Killing form is negative semi-definite. Since in fact the Killing form has signature $(2, 1)$, $\mathfrak{sl}(2, \mathbb{R})$ is not a Lie subalgebra of $\mathfrak{so}(m)$ for any m. Let σ be a representation of $\mathrm{SL}(2, \mathbb{R})$ to $\mathrm{SO}(m)$ for some m. Let $\tilde{f}_i = \sigma_* f_i$. The $\tilde{f}_i$ satisfy the bracket relations of the f_i since σ_* is a Lie algebra homomorphism. Since σ_* is not injective, we have a non-trivial relation of the form

$$x := a_1\tilde{f}_1 + a_2\tilde{f}_2 + a_3\tilde{f}_3 = 0\,.$$

Suppose $a_1 \neq 0$. We have $\operatorname{ad}(\tilde{f}_3)\operatorname{ad}(\tilde{f}_2)x = -4a_1\tilde{f}_1$. Therefore, $\tilde{f}_1 = 0$. Consequently,

$$\operatorname{range}\{\operatorname{ad}(\tilde{f}_1)\} = 0 \quad \text{so} \quad \tilde{f}_2 = 0 \quad \text{and} \quad \tilde{f}_3 = 0\,.$$

Therefore, σ_* is the zero map. The argument is the same with the remaining coefficients. Since σ_* is the zero map, σ is the trivial representation by Lemma 6.2. This proves Assertion 2.

Let $\xi = a^1 f_1 + a^2 f_2 + a^3 f_3 \in \mathfrak{g}$. Then $\xi^2 = \langle \xi, \xi \rangle \operatorname{Id}$. Set $\lambda = \langle \xi, \xi \rangle$. Note that $\xi^3 = \lambda\xi$, $\lambda^4 = \lambda^2$ and so forth. This yields the following consequences.

1. If $\lambda < 0$, let $\mu = \sqrt{-\lambda}$. Then:

$$\begin{aligned} \exp(\xi) &= (1 + \tfrac{1}{2!}\lambda + \tfrac{1}{4!}\lambda^2 + \cdots)\operatorname{Id} + (1 + \tfrac{1}{3!}\lambda + \tfrac{1}{5!}\lambda^2 + \cdots)\xi \\ &= (1 - \tfrac{1}{2!}\mu^2 + \tfrac{1}{4!}\mu^4 + \cdots)\operatorname{Id} + \mu^{-1}(\mu - \tfrac{1}{3!}\mu^3 + \tfrac{1}{5!}\mu^5 + \cdots)\xi \\ &= \cos(\mu)\operatorname{Id} + \mu^{-1}\sin(\mu)\xi\,. \end{aligned}$$

2. If $\lambda = 0$, then $\exp(\xi) = \operatorname{Id} + \xi$.

3. If $\lambda > 0$, let $\mu = \sqrt{\lambda}$. Then:

$$\begin{aligned} \exp(\xi) &= (1 + \tfrac{1}{2!}\lambda + \tfrac{1}{4!}\lambda^2 + \cdots)\operatorname{Id} + (1 + \tfrac{1}{3!}\lambda + \tfrac{1}{5!}\lambda^2 + \cdots)\xi \\ &= (1 + \tfrac{1}{2!}\mu^2 + \tfrac{1}{4!}\mu^4 + \cdots)\operatorname{Id} + \mu^{-1}(\mu + \tfrac{1}{3!}\mu^3 + \tfrac{1}{5!}\mu^5 + \cdots)\xi \\ &= \cosh(\mu)\operatorname{Id} + \mu^{-1}\sinh(\mu)\xi\,. \end{aligned}$$

This shows $\exp(\xi) \in \operatorname{span}\{\operatorname{Id}, \xi\}$. Let

$$g := -\operatorname{Id} + f_1 + f_2 = \begin{pmatrix} -1 & 2 \\ 0 & -1 \end{pmatrix}.$$

As $f_1 + f_2$ is a null vector, $\exp(t\xi) = \operatorname{Id} + t\xi \neq -\operatorname{Id}$ for any t so $g \notin \operatorname{range}\{\exp^{\mathfrak{g}}\}$. This proves Assertion 3.

By Lemma 6.11, $\pi_1(\mathrm{SL}(2,\mathbb{R})) = \pi_1(\mathrm{SO}(2)) = \pi_1(S^1) = \mathbb{Z}$. This also follows as we can identify $\mathrm{SL}(2,\mathbb{R})$ with the pseudo-sphere

$$(x^0)^2 + (x^1)^2 - (x^2)^2 - (x^3)^2 = 1$$

and this deformation retracts onto the circle $(x^0)^2 + (x^1)^2 = 1$, $x^2 = x^3 = 0$. We now use the fact that $\mathfrak{sl}(2,\mathbb{R}) \otimes_{\mathbb{R}} \mathbb{C} = \mathfrak{su}(2) \otimes_{\mathbb{R}} \mathbb{C}$ which was observed above. Let π be the covering projection from $\overline{\mathrm{SL}(2,\mathbb{R})}$ to $\mathrm{SL}(2,\mathbb{R})$. We use π to identify the Lie algebra of $\overline{\mathrm{SL}(2,\mathbb{R})}$ with $\mathfrak{sl}(2,\mathbb{R})$. Let Θ be a representation of $\overline{\mathrm{SL}(2,\mathbb{R})}$ to $\mathrm{GL}(m,\mathbb{R})$ for some m and let θ be the associated Lie algebra homomorphism from $\mathfrak{sl}(2,\mathbb{R})$ to $\mathfrak{gl}(2,\mathbb{R})$. We complexify to regard θ_c as a Lie algebra homomorphism from $\mathfrak{sl} \otimes_{\mathbb{R}} \mathbb{C}$ to $\mathfrak{gl}(m,\mathbb{C})$. We may then restrict to regard θ_c as a Lie algebra homomorphism $\theta_{\mathfrak{su}(2)}$ from $\mathfrak{su}(2)$ to $\mathfrak{gl}(m,\mathbb{C})$. Since $\mathrm{SU}(2)$ is compact and simply connected, $\mathfrak{su}(2)$ arises from a representation $\Theta_{\mathrm{SU}(2)}$ from $\mathrm{SU}(2)$ to $\mathrm{GL}(2,\mathbb{C})$.

By the Peter–Weyl Theorem, every finite-dimensional representation of S^3 is unitarily equivalent to a subrepresentation of L^2. We apply Theorem 5.12 to find an orthogonal direct sum decomposition of $L^2(S^3) = \oplus_{j\in\mathbb{N}} H(4, j)$ where $H(4, j)$ is the subspace of homogeneous polynomials of degree j which are harmonic with respect to the Euclidean Laplacian on $\mathbb{R}^4$. The decomposition of L^2 is equivariant with respect to the isometry group of S^3. Consequently, every finite-dimensional representation of S^3 is equivalent to a subrepresentation of $\oplus_{j\le j_0} H(4, j)$. In particular, it is polynomial in the standard coordinates on $\mathbb{R}^4$, i.e., in the coordinates x^i. Therefore, it extends to a map $\Theta_{\mathrm{SU}(2)}$ from SU(2) to $\mathrm{GL}(2,\mathbb{C})$. Since $\theta_c(x\cdot\tilde{x}) = \theta_c(x)\cdot\theta_c(\tilde{x})$, the same identity holds for complex coordinates and, consequently, Θ_c is a representation from $\mathrm{SU}_{\mathbb{C}}(2)$ to $\mathrm{GL}(2,\mathbb{C})$. Restricting to $\mathrm{SL}(2,\mathbb{R})$, we see that Θ in fact arises from a representation of $\mathrm{SL}(2,\mathbb{R})$. Consequently, $\ker\{\pi\}\subset\ker\{\Theta\}$ and Θ is not a faithful representation of $\overline{\mathrm{SL}(2,\mathbb{R})}$. This shows that $\overline{\mathrm{SL}(2,\mathbb{R})}$ is not a matrix group. □

6.9 THE COHOMOLOGY OF COMPACT LIE GROUPS

Let $\mathfrak{g}$ be the Lie algebra of a compact connected Lie group; $\mathfrak{g}$ is vector space of left-invariant vector fields on G. Let $[\cdot,\cdot]$ be the Lie bracket. Let $\mathfrak{g}^*$ be dual space; this is the vector space of left-invariant 1-forms. The exterior derivative and the Lie bracket are related by the formula:

$$d\omega[x, y] = -[x, y](\omega)\,.$$

Let $\{e_i\}$ be a basis for $\mathfrak{g}$ and let $\{e^i\}$ be the corresponding dual basis for $\mathfrak{g}^*$. We expand

$$[e_i, e_j] = c_{ij}{}^k e_k$$

to define the *Lie algebra structure constants* of Equation (6.2.b). We have:

$$de^k = -c_{ij}{}^k e^i \wedge e^j\,.$$

We consider the finite cochain complex $\{\Lambda^p(\mathfrak{g}^*), d\}$ and let $H^*(\mathfrak{g}^*, d)$ be the associated cohomology. The following result of Chevalley and Eilenberg [14] provides a method for computing the de Rham cohomology of G using Lie theoretic methods. The proof we give is based on the Hodge Decomposition Theorem (see Theorem 5.13); it is also possible to proceed directly.

Theorem 6.26 *Let G be a compact connected Lie group which is equipped with a bi-invariant metric of total volume* 1.

1. *If $\omega_p\in\ker\{\Delta^p\}$, then ω_p is bi-invariant.*
2. *If $\omega_p\in\Lambda^p(\mathfrak{g}^*)$ and if $\omega_p = d\psi_{p-1}$ for some ψ_{p-1} in $C^\infty(\Lambda^{p-1}G)$, then there exists Ψ_{p-1} in $\Lambda^{p-1}(\mathfrak{g}^*)$ so that $\omega_p = d\Psi_{p-1}$.*
3. *The inclusion map $i : (\Lambda^p(\mathfrak{g}^*), d)\to(C^\infty(\Lambda^p G), d)$ is a chain map which defines an isomorphism from $H^p(\mathfrak{g}^*, d)$ to $H^p_{\mathrm{dR}}(G)$.*

Proof. Let ω be a p-form with $\Delta^p = 0$, and let $h \in G$. Because the metric is left-invariant, $L_h^*\Delta = \Delta L_h^*$. Therefore, $L_h^*\omega \in \ker\{\Delta^p\}$ as well. On the other hand, G is assumed to be connected. Consequently, there is a smooth curve $\gamma(t)$ connecting h to the identity and providing a homotopy between L_h and L_{Id}. The homotopy axiom then shows $[L_h^*\omega] = [\omega]$ in de Rham cohomology. The Hodge Decomposition Theorem (see Theorem 5.13) then shows $L_h^*\omega = \omega$. The argument is the same for right-invariance; Assertion 1 now follows.

Suppose ω is left-invariant and that $\omega = d\psi$ for some smooth $(p-1)$-form which is not necessarily left-invariant. We average over the group to see that:

$$\begin{aligned} \omega &= \int_G (L_h^*\omega)\,\mathrm{dvol}(h) = \int_G (L_h^* d\psi)\,\mathrm{dvol}(h) = d\int_G (L_h^*\psi)\,\mathrm{dvol}(h) \\ &= d\{\Psi\} \quad \text{where} \quad \Psi := \int_G (L_h^*\psi)\,\mathrm{dvol}(h)\,. \end{aligned}$$

Assertion 2 now follows by noting Ψ is left-invariant. We have used, of course, the fact that we can interchange the order of differentiation and integration. Assertion 1 shows that $[i]$ is surjective and Assertion 2 shows that $[i]$ is injective as a map from $H^p(\mathfrak{g}^*, d)$ to $H^p_{\mathrm{dR}}(G)$. Assertion 3 now follows. □

6.9.1 THE HOPF STRUCTURE THEOREM. The results in this section arise from work of H. Hopf [33].

H. Hopf (1894–1971)

We refer the reader to the discussion in Section 8.1.6 for the definition of a unital graded ring. We say that a connected unital graded ring R is a *co-ring* if we have a co-multiplication θ which is a graded ring morphism from R to $R \otimes R$ which is *co-associative*, i.e.,

$$(\mathrm{Id} \otimes \theta) \circ \theta = (\theta \otimes \mathrm{Id}) \circ \theta\,.$$

We do not assume the co-multiplication is co-commutative. We say that R is a *Hopf algebra* if, in addition, there is an augmentation $\varepsilon : R \to F$ so

$$(\mathrm{Id} \otimes \epsilon) \circ \theta = \mathrm{Id} \quad \text{and} \quad (\epsilon \otimes \mathrm{Id}) \circ \theta = \mathrm{Id}\,.$$

We use the existence of a co-unit to pin θ down slightly. We expand

$$\theta(a) = 1 \otimes b_n + \text{stuff} + a_n \otimes 1$$

where the deleted material "stuff" belongs to $\oplus_{i>0, j>0, i+j=n} R_i \otimes R_j$. We conclude

$$\theta(a) = a \otimes 1 + \text{stuff} + 1 \otimes a\,.$$

We have the following examples of Hopf algebras.

1. Let $\mathbb{R}[x_{2\nu}] = \mathbb{R} \oplus \mathbb{R} \cdot x_{2\nu} \oplus \mathbb{R} \cdot x_{2\nu}^2 \oplus \cdots$ be the polynomial ring on an indeterminate $x_{2\nu}$ of even degree $2\nu > 0$. Define $\theta(x) = x \otimes 1 + 1 \otimes x$; this is then a Hopf algebra.

2. Let $\Lambda[y_{2\nu+1}] = \mathbb{R} \oplus \mathbb{R} \cdot y_{2\nu+1}$ be the exterior algebra on a generator $y_{2\nu+1}$ of odd degree $2\nu + 1 \geq 1$.

3. The tensor product of Hopf algebras is again a Hopf algebra. Therefore,
$$R := \mathbb{R}[x_{2\nu_1}, \ldots, x_{2\nu_k}] \otimes \Lambda[y_{2\sigma_1+1}, \ldots, y_{2\sigma_\ell+1}]$$
is a finitely generated Hopf algebra.

Lemma 6.27 Let B be a Hopf algebra and let A be a Hopf subalgebra. Assume that A and some other element $x \notin A$ of positive degree generate B as an algebra. Then:
$$B = \left\{ \begin{array}{lll} \Lambda[x] & \text{if} & \deg(x) \text{ is } odd \\ \mathbb{R}[x] & \text{if} & \deg(x) \text{ is } even \end{array} \right\} \otimes_{\mathbb{R}} A\,.$$

Proof. Let $A^+ = \oplus_{j>0} A_j$ be the elements of positive order. Let $\mathcal{I} := \sum x^\nu a_\nu$ for $a_\nu \in A$ and $\deg(a_\nu) > 0$. This is an ideal of B and $x \notin \mathcal{I}$. Let
$$\pi : B \to B/\mathcal{I} \quad \text{and} \quad \psi = (\mathrm{Id} \otimes \pi) \circ \theta : B \to B \otimes B/\mathcal{I}\,.$$
We may decompose $\psi(x) = x \otimes 1 + \text{stuff} + 1 \otimes \pi(x)$. Since the "stuff" lies in $B_i \otimes B_j$ for $i > 1$, $j > 1$, and $i + j = \deg(x)$, these elements must belong to A^+ and, consequently, these elements are all killed by π. Consequently, $\psi(x) = x \otimes 1 + 1 \otimes \pi(x)$. Because A is a Hopf subalgebra, $\psi(a) = a \otimes 1$ if $a \in A^+$. Let $\deg(x) = n$. Suppose we had a relation of the form $a_1 + a_2 x = 0$ for the $a_i \in A$ and $a_2 \neq 0$. Then $a_i \in A^+$ since $x \notin A$. We compute
$$\begin{aligned} 0 &= \psi(a_1 + a_2 x) = \psi(a_1) + \psi(a_2)\psi(x) \\ &= (a_1 + a_2 x) \otimes 1 + a_2 \otimes \pi(x) = a_2 \otimes \pi(x)\,. \end{aligned}$$
This implies $a_2 = 0$ since $\pi(x) \neq 0$. If n is odd, $x^2 = 0$ so $B = A \oplus A \cdot x$ and we are done. Therefore, we may suppose that n is even. Suppose we have a non-trivial relation
$$a_0 + \cdots + a_\nu x^\nu = 0\,.$$
Choose ν minimal with $\nu \geq 2$. Apply ψ to see $\psi(a_\nu) = a_\nu \otimes 1$. If $\deg(a_\nu) = 0$, then a_ν is a scalar and this is automatic. If $\deg(a_\nu) > 0$, this also is automatic.
$$\sum_\nu a_\nu (x \otimes 1 + 1 \otimes \pi(x))^\nu = 0\,.$$
We compare graded degrees to see $\sum_\nu \nu \cdot a_\nu x^{\nu-1} \otimes \pi(x) = 0$. Fortunately we are in characteristic zero. This gives a relation of lower degree. □

Theorem 6.28 *If R is a finitely generated Hopf algebra, then R is isomorphic as a graded associative ring to $\mathbb{R}[x_{2\nu_1}, \dots, x_{2\nu_k}] \otimes \Lambda[y_{2\sigma_1+1}, \dots, y_{2\sigma_\ell+1}]$. If R is finite-dimensional, then there is no polynomial component.*

Proof. We suppose B is a finitely generated Hopf algebra. We choose generators

$$\{x_1, \dots, x_n\} \quad \text{so} \quad \deg(x_i) \leq \deg(x_{i+1}).$$

Let B_ν be the subalgebra generated by $\{x_1, \dots, x_\nu\}$. Then $B_{\nu+1}$ is generated by B_ν and $x_{\nu+1}$. Furthermore, $\theta(x_\nu) = x_\nu \otimes 1 + \text{stuff} + 1 \otimes x_\nu$ so counting degrees the stuff belongs to B_ν. Therefore, B_ν is a Hopf algebra. We apply Lemma 6.27 and induction. □

We use Theorem 6.28 to see:

Theorem 6.29 *If G is a compact connected Lie group, then $H^*_{\mathrm{dR}}(G)$ is isomorphic to an exterior algebra $\Lambda^*[x_{2i_1-1}, \dots, x_{2i_\ell-1}]$ on a finite number of odd-dimensional generators.*

Proof. Let G be a compact connected Lie group. By Theorem 5.7, $H^p_{\mathrm{dR}}(G)$ is finite-dimensional for $0 \leq p \leq m$; $H^p_{\mathrm{dR}}(G) = 0$ for $p > m$. Let $m : G \times G \to G$ be the multiplication in the group. This induces a ring homomorphism $m^* : H^*_{\mathrm{dR}}(G) \to H^*_{\mathrm{dR}}(G \times G)$. By the Künneth formula, we have a natural graded ring isomorphism

$$\pi_1^* \wedge \pi_2^* : H^*_{\mathrm{dR}}(G) \otimes H^*_{\mathrm{dR}}(G) \to H^*_{\mathrm{dR}}(G \times G).$$

Thus, $\theta := (\pi_1^* \wedge \pi_2^*)^{-1} \circ m^*$ provides a co-multiplication from $H^*_{\mathrm{dR}}(G)$ to $H^*_{\mathrm{dR}}(G \times G)$. Let ϵ be the inclusion of the identity of the group into G. Dually, this gives a map ϵ^* from $H^*_{\mathrm{dR}}(G)$ to $H^0_{\mathrm{dR}}(G) = \mathbb{R}$. A bit of diagram chasing shows that H^*_{dR} is a Hopf algebra and the desired conclusion now follows from Theorem 5.16. □

6.10 THE COHOMOLOGY OF THE UNITARY GROUP

We illustrate Theorem 5.18 by taking $G = \mathrm{U}(n)$ to be the complex unitary group. The *Maurer–Cartan form* [8] $\Theta := g^{-1}dg$ plays a crucial role.

Theorem 6.30 *Set $\omega_{2k-1} := \mathrm{Tr}\{(g^{-1}dg)^{2k-1}\}$. Then ω_{2k-1} is a closed $(2k-1)$-form on $\mathrm{U}(n)$. If $x_{2k-1} := [\omega_{2k-1}]$ is the corresponding cohomology class, then*

$$H^*_{\mathrm{dR}}(\mathrm{U}(n)) \otimes_{\mathbb{R}} \mathbb{C} = \Lambda_{\mathbb{C}}[\omega_1, \dots, \omega_{2n-1}].$$

Proof. We first show $d\omega_{2k-1} = 0$. Let $\Theta := g^{-1}dg$. Then

$$\begin{aligned}
d\Theta &= d(g^{-1}) \cdot dg = -g^{-1}dg \cdot g^{-1}dg = -\Theta^2, \\
d\omega_{2k-1} &= d(\mathrm{Tr}\{\Theta^{2k-1}\}) = \mathrm{Tr}\{d(\Theta^{2k-1})\} \\
&= \mathrm{Tr}\{d\Theta \wedge \Theta^{2k-2} - \Theta \wedge d\Theta \wedge \Theta^{2k-3} + \cdots\} \\
&= -\mathrm{Tr}\{\Theta^{2k}\} = -\mathrm{Tr}\{\Theta \cdot \Theta^{2k-1}\} = \mathrm{Tr}\{\Theta^{2k-1} \cdot \Theta\} = \mathrm{Tr}\{\Theta^{2k}\}.
\end{aligned}$$

It now follows that $d\omega_{2k-1} = -d\omega_{2k-1}$ and hence $d\omega_{2k-1} = 0$. We are using the fact, of course, that Θ is a matrix of 1-forms. The 1-forms anti-commute and the trace commutes. Let i_{n-1} be the inclusion of $\mathrm{U}(n-1)$ in $\mathrm{U}(n)$. We have

$$i_{n-1}(T) = \begin{pmatrix} T & 0 \\ 0 & \mathrm{Id} \end{pmatrix}, \quad g_n^{-1} \cdot dg_n(i_{n-1}(T)) = \begin{pmatrix} g_{n-1}^{-1} dg_{n-1} & 0 \\ 0 & 0 \end{pmatrix},$$
$$i_{n-1}(\Theta_n^{2k-1}) = \begin{pmatrix} \Theta_{n-1}^{2k-1} & 0 \\ 0 & 0 \end{pmatrix}, \quad i_{n-1}(\mathrm{Tr}\{\Theta_n^{2k-1}\}) = \mathrm{Tr}\{\Theta_{n-1}^{2k-1}\}.$$

Therefore, these classes are universal and we do not need to indicate the dimension n explicitly. By Lemma 5.10, there exists $g : S^{2n-1} \to \mathrm{U}(2^n)$ so that

$$\int_{S^{2n-1}} g^* \omega_{2n-1} \neq 0.$$

The fibration $\mathrm{U}(\ell-1) \to \mathrm{U}(\ell) \to S^{2\ell-1}$ yields a sequence:

$$0 \to \pi_{2n-1}(\mathrm{U}(\ell-1)) \to \pi_{2n-1}(\mathrm{U}(\ell)) \to 0 \quad \text{for} \quad \ell > n.$$

Consequently, the map $\pi_{2k-1}(\mathrm{U}(k)) \to \pi_{2k-1}(\mathrm{U}(2^k))$ is injective. Therefore, in fact, after "smoothing" we can push g down to a map from S^{2n-1} to $\mathrm{U}(n)$. This yields maps f_ν from $S^{2\nu-1}$ to $\mathrm{U}(k)$ for $1 \leq \nu \leq k$ so that

$$\int_{S^{2\nu-1}} f_\nu^* \omega_{2\nu-1} \neq 0.$$

So none of these classes is zero in $H_{\mathrm{dR}}^{2\nu-1}$. Theorem 6.29 shows

$$H_{\mathrm{dR}}^\star(\mathrm{U}(k)) = \Lambda(y_{2\mu_1-1}, \ldots, y_{2\mu_\ell-1}) \quad \text{where} \quad 1 \leq \mu_1 \leq \cdots \leq \mu_\ell.$$

If $\mu_i \leq \nu$, then $f_\nu^* y_{2\mu_i-1} = 0$ in $H_{\mathrm{dR}}^*(S^{2\nu-1})$. Consequently,

$$f_\nu^*\{\Lambda[y_{2\mu_1-1}, \ldots, y_{2\mu_i-1}]\} = 0 \quad \text{so} \quad [\omega_{2\nu-1}] \notin \Lambda[y_{2\mu_1-1}, \ldots, y_{2\mu_i-1}] \quad \text{if} \quad \mu_i \leq \nu.$$

Consequently, we may assume that $[\omega_{2\nu-1}]$ is taken as one of the generators. Since

$$0 \neq y_{2\mu_1-1} \wedge \cdots \wedge y_{2\mu_\ell-1} \quad \text{in} \quad H_{\mathrm{dR}}^*(\mathrm{U}(k)),$$

$2\mu_1 - 1 + \cdots + 2\mu_\ell - 1 \leq \dim\{\mathrm{U}(k)\}$. Since $1 + 3 + \cdots + 2k - 1 = \dim\{\mathrm{U}(k)\}$, there are no other generators than the $[\omega_{2\nu-1}]$. Consequently, as desired,

$$H_{\mathrm{dR}}^*(\mathrm{U}(k)) = \Lambda([\omega_1], \ldots, [\omega_{2k-1}]).$$

□

CHAPTER 7

Homogeneous Spaces and Symmetric Spaces

Good auxiliary references for the material of Chapter 7 are Arvanitoyeorgos [2], Helgason [30], O'Neill [49], Warner [60], and Ziller [64]. In Section 7.1, we prove results about smooth structures on coset spaces. In Section 7.2, we discuss the isometry group of a pseudo-Riemannian manifold. In Section 7.3, we treat Killing vector fields. In Section 7.4, we establish facts concerning homogeneous pseudo-Riemannian manifolds. In Section 7.5, we use Jacobi vector fields to study the geometry of local symmetric spaces. In Section 7.6, we examine symmetric spaces.

7.1 SMOOTH STRUCTURES ON COSET SPACES

Let H be a Lie group which acts continuously on a topological space X from the right. If $x \in X$, let $[x] := x \cdot H$ and let $\pi : x \to [x]$ be the natural projection from X to the set of right cosets X/H. Give X/H the quotient topology; a subset $\mathcal{U}$ of X/H is open if and only if $\pi^{-1}(\mathcal{U})$ is open in X. It is immediate that π is continuous; the group action can be used to show that π is an open map. If $f : X \to Y$ satisfies $f(x \cdot h) = f(x)$ for all $x \in X$ and all $h \in H$, let $[f] : X/H \to Y$ be the induced map on the coset space; $[f] : X/H \to Y$ is continuous if and only if $f : X \to Y$ is continuous.

We say that $F \to E \xrightarrow{\pi} M$ is a *fiber bundle* if π is a smooth map from a manifold E onto a manifold M, and if there is a cover of M by open sets $\mathcal{O}_\alpha$ such that there exist fiber preserving diffeomorphisms from $\pi^{-1}(\mathcal{O}_\alpha)$ to $\mathcal{O}_\alpha \times F$. We may identify F with $\pi^{-1}(P)$ for any $P \in M$. The fiber bundle is said to be a *principal bundle* if F is a Lie group which acts smoothly on E without fixed points from the right and if $M = E/F$ is the associated coset space; see the discussion in Section 2.1.9 of Book I for further details.

Theorem 7.1 *Let H be a closed subgroup of a Lie group G.*

1. *G/H has a natural smooth structure so that $H \to G \xrightarrow{\pi} G/H$ is a principal H-bundle and so that the left action of G on G/H is smooth.*

2. *If H is a normal subgroup of G, then the coset space G/H is a Lie group.*

3. *If ρ is a group homomorphism from G onto a Lie group $\tilde{G}$ so that $\rho(H)$ is a Lie subgroup of $\tilde{G}$, then ρ induces a smooth map from G/H onto $\tilde{G}/\rho(H)$.*

Proof. If $[x] = [y]$ in G/H, then $[L_g x] = [L_g y]$. Let $[L_g][x] := [L_g x]$ define a continuous left action of G on G/H. Let $\mathfrak{h}$ and $\mathfrak{g}$ be the Lie algebras of H and of G, respectively. Use a right-invariant Riemannian metric on G to decompose $\mathfrak{g} = \mathfrak{h} \oplus \mathfrak{h}^\perp$. Let $B_{\epsilon,\mathfrak{h}}$ (resp. $B_{\epsilon,\mathfrak{h}^\perp}$) be the open ball of radius $\epsilon > 0$ about the origin in $\mathfrak{h}$ (resp. $\mathfrak{h}^\perp$). Let

$$\Xi(\eta_{\mathfrak{h}}, \eta_{\mathfrak{h}^\perp}) := \exp(\eta_{\mathfrak{h}^\perp}) \cdot \exp(\eta_{\mathfrak{h}}) \quad \text{for} \quad \eta_{\mathfrak{h}} \in B_{\epsilon,\mathfrak{h}} \quad \text{and} \quad \eta_{\mathfrak{h}^\perp} \in B_{\epsilon,\mathfrak{h}^\perp}\,.$$

Then Ξ is a diffeomorphism from $B_{\epsilon,\mathfrak{h}} \times B_{\epsilon,\mathfrak{h}^\perp}$ to a neighborhood of the identity in G if ϵ is sufficiently small. Let $\mathcal{O}$ be an open subset of $B_{\epsilon,\mathfrak{h}^\perp}$. The following are open sets of G:

$$\begin{aligned}&\exp(\mathcal{O}) \cdot \exp(B_{\epsilon,\mathfrak{h}}) = \Xi(B_{\epsilon,\mathfrak{h}} \times \mathcal{O}), \quad \exp(\mathcal{O}) \cdot H = \exp(\mathcal{O}) \cdot \exp(B_{\epsilon,\mathfrak{h}}) \cdot H, \\ &\pi^{-1}\{\pi(\exp(\mathcal{O}))\} = \exp(\mathcal{O}) \cdot H\,.\end{aligned}$$

Therefore, $\pi^{-1}\{\pi(\exp(\mathcal{O}))\}$ is an open subset of G so $\pi \circ \exp$ is a continuous open map from $B_{\epsilon,\mathfrak{h}^\perp}$ onto the open neighborhood $\tilde{B}_\epsilon := (\pi \circ \exp)(B_{\epsilon,\mathfrak{h}^\perp})$ of [Id]. If

$$(\pi \circ \exp)(\eta_{1,\mathfrak{h}^\perp}) = (\pi \circ \exp)(\eta_{2,\mathfrak{h}^\perp}) \quad \text{for} \quad \eta_{i,\mathfrak{h}^\perp} \in B_{\epsilon,\mathfrak{h}^\perp}\,,$$

then $\exp(\eta_{1,\mathfrak{h}^\perp}) = \exp(\eta_{2,\mathfrak{h}^\perp}) \cdot h$ for some $h \in H$. If ϵ is sufficiently small, then $h = \exp(\eta_{\mathfrak{h}})$ for $\eta_{\mathfrak{h}}$ close to 0 and $\exp(\eta_{1,\mathfrak{h}^\perp}) = (\eta_{2,\mathfrak{h}^\perp}) \cdot \exp(\eta_{\mathfrak{h}})$ for some $h \in \mathfrak{h}$ close to 0. This implies

$$\Xi(0, \eta_{1,\mathfrak{h}^\perp}) = \Xi(\eta_{\mathfrak{h}}, \eta_{2,\mathfrak{h}^\perp})$$

and, therefore, $\eta_{1,\mathfrak{h}^\perp} = \eta_{2,\mathfrak{h}^\perp}$. Consequently, $\pi \circ \exp$ is a homeomorphism from $B_{\epsilon,\mathfrak{h}^\perp}$ to $\tilde{B}_\epsilon$ which defines a coordinate chart near [Id]. If $[x] \in \tilde{B}_\epsilon$, let

$$\eta_{\mathfrak{h}^\perp} : [x] \to (\pi \circ \exp)^{-1}([x]) \in B_{\epsilon,\mathfrak{h}^\perp} \tag{7.1.a}$$

be the inverse homeomorphism. We define a continuous section s to π over $\tilde{B}_\epsilon$ by setting:

$$s([x]) := \exp(\eta_{\mathfrak{h}^\perp}([x])) : \tilde{B}_\epsilon \to \exp(B_{\epsilon,\mathfrak{h}}) \subset G\,. \tag{7.1.b}$$

We then have that $\pi(s([x])) = [x]$ for $[x] \in \tilde{B}_\epsilon$. We use $[L_g]$ to translate the chart $\tilde{B}_\epsilon$ and thereby obtain charts covering G/H as $g \in G$. If $\pi_{\mathfrak{h}^\perp}$ is orthogonal projection on $\mathfrak{h}^\perp$, then the transition function relating the given chart to the translated chart is given by the smooth function

$$\eta_{\mathfrak{h}^\perp} \to \pi_{\mathfrak{h}^\perp}(\Xi^{-1}(L_g(\exp(\eta_{\mathfrak{h}^\perp}))))\,.$$

Therefore, G/H admits a smooth manifold structure, $[L_g]$ is a smooth left action of G on G/H, and s is a smooth section to π over $\tilde{B}_\epsilon$. If $y \in \pi^{-1}(\tilde{B}_\epsilon) = B_{\epsilon,\mathfrak{h}^\perp} \times H$, then we may express

$$y = s([y]) \cdot h(y) \quad \text{where} \quad h(y) := s([y])^{-1} \cdot y\,.$$

The map $y \to ([y], h(y))$ is an H-equivariant diffeomorphism from $\pi^{-1}(\tilde{B}_\epsilon)$ onto $\tilde{B}_\epsilon \times H$; the inverse map is $([y], h) \to s([y]) \cdot h$. Consequently, $H \to G \xrightarrow{\pi} G/H$ is a principal H-bundle. This proves Assertion 1.

If H is a normal subgroup of G, then G/H is a group. Let s_i be local smooth sections to π which are defined near $[x_i]$. As

$$[x_1] * [x_2] = [s([x_1]) * s([x_2])] \quad \text{and} \quad [x_1]^{-1} = [s([x_1])^{-1}],$$

the group multiplication and inversion are smooth. Therefore, G/H is a Lie group; this proves Assertion 2. Assertion 3 is immediate from the discussion given above. □

Let G be a Lie group which acts smoothly and transitively on a manifold M from the left. Fix $P \in M$. Let $H = H_P := \{g \in G : g \cdot P = P\}$ be the isotropy subgroup; H is a closed subgroup of G. Let $\theta : g \to g \cdot P$; θ is a smooth map from G onto M. If $h \in H$, then

$$\theta(g \cdot h) = g \cdot h \cdot P = g \cdot P = \theta(g)$$

so θ induces a continuous map $[\theta] : G/H$ onto M. We show that $[\theta]$ is 1-1 as follows:

$$\begin{aligned} \theta(g_1) = \theta(g_2) \quad &\Leftrightarrow \quad g_1 \cdot P = g_2 \cdot P \quad \Leftrightarrow \quad g_2^{-1} g_1 \cdot P = P \\ &\Leftrightarrow \quad g_2^{-1} g_1 \in H \quad \Leftrightarrow \quad g_1 \in g_2 H \quad \Leftrightarrow \quad [g_1] = [g_2] \ \text{in} \ G/H. \end{aligned}$$

Theorem 7.2 *Let G be a Lie group which acts smoothly and transitively on a manifold M from the left. Then $[\theta]$ is a G-equivariant diffeomorphism from G/H to M.*

Proof. Since $[\theta]$ is G-equivariant, it suffices to show that $[\theta]$ is a diffeomorphism near [Id]. We adopt the notation established in the proof of Theorem 7.1. Let $\eta_{\mathfrak{h}^\perp} : \tilde{B}_\epsilon \to B_{\epsilon,\mathfrak{h}^\perp}$ be the diffeomorphism of Equation (7.1.a) and let $s([x]) = \exp(\eta_{\mathfrak{h}^\perp}([x]))$ be the section of Equation (7.1.b). Since $[\theta][x] = s(x) \cdot P$, $[\theta]$ is smooth. For $0 \neq \eta_{\mathfrak{h}^\perp} \in B_{\epsilon,\mathfrak{h}^\perp}$, let

$$\Psi(t, Q; \eta_{\mathfrak{h}^\perp}) := \exp(t\eta_{\mathfrak{h}^\perp}) \cdot Q$$

define a 1-parameter flow on M with associated vector field $X_{\eta_{\mathfrak{h}^\perp}}$. We have

$$X_{\eta_{\mathfrak{h}^\perp}}(P) = \partial_t\{\exp(t\eta_{\mathfrak{h}^\perp}) \cdot P\}|_{t=0} = ([\theta]_* \circ \pi_* \circ \exp_*)\{\eta_{\mathfrak{h}^\perp}\}.$$

If $X_{\eta_{\mathfrak{h}^\perp}}(P) = 0$, then the Fundamental Theorem of ODEs shows $\Psi(t, P; \eta_{\mathfrak{h}^\perp}) = P$ for all t. However, $\Psi(t, P; \eta_{\mathfrak{h}^\perp}) = ([\theta] \circ \pi \circ \exp)(t\eta_{\mathfrak{h}^\perp}) \neq P$ if $t \neq 0$ since $[\theta]$ is 1-1 from $\tilde{B}_\epsilon$ to M and since $\pi \circ \exp$ is a diffeomorphism from $B_{\epsilon,\mathfrak{h}^\perp}$ to $\tilde{B}_\epsilon$. Consequently, $\{[\theta] \circ \pi \circ \exp\}_*$ is an injective map from $T_0 B_{\epsilon,\mathfrak{h}^\perp}$ to $T_P M$ and hence $[\theta]$ is a diffeomorphism from a neighborhood of [Id] in G/H to a neighborhood of P in M. Using left translation in the group, we see the same is true for any element of G/H and, therefore, $[\theta]$ is a global diffeomorphism. □

Theorem 7.3 *Let A and B be closed subgroups of a Lie group C with $A \subset B \subset C$. Then the following sequence is a smooth fiber bundle:*

$$B/A \to C/A \stackrel{\pi_{C/A \to C/B}}{\longrightarrow} C/B\,.$$

Proof. Let $x \in C$. By Theorem 7.1, there exists an open neighborhood $\mathcal{O}$ of the coset $[x]_{C/B}$ in C/B and a smooth section s to the fibration $B \to C \stackrel{\pi_{C \to C/B}}{\longrightarrow} C/B$ defined on $\mathcal{O}$. The maps

$$\Xi : ([x]_{C/B}, b) \to s([x]_{C/B}) \cdot b \quad \text{and} \quad \Xi^{-1} : x \to ([x]_{C/B}, s([x]_{C/B})^{-1} x)$$

are diffeomorphisms between $\mathcal{O} \times B$ and $\pi^{-1}_{C \to C/B}(\mathcal{O})$ in C. The maps

$$\Xi_1 : ([x]_{C/B}, [b]_{B/A}) \to \left[s([x]_{C/B}) \cdot [b]_{B/A}\right]_{C/A}, \quad \text{and}$$

$$\Xi_1^{-1} : [x]_{C/A} \to \left(\pi_{C/A \to C/B}([x]_{C/A}), s(\pi_{C/A \to C/B}([x]_{C/A}))^{-1} \cdot \pi_{C/A \to C/B}([x]_{C/A})\right)$$

are diffeomorphisms between $\mathcal{O} \times [B/A]$ and $\pi^{-1}_{C/A \to C/B}(\mathcal{O})$ in C/A that provide the required local trivialization of the natural map from C/B to C/A. □

We have the following examples.

The special orthogonal group $\mathrm{SO}(m+1)$ acts transitively on the unit sphere S^m in $\mathbb{R}^{m+1}$ by isometries with isotropy subgroup $\mathrm{SO}(m)$; $\mathrm{SO}(m) \to \mathrm{SO}(m+1) \to S^m$ is a principle $\mathrm{SO}(m)$ bundle.

The unitary group $\mathrm{U}(m)$ acts transitively on the unit sphere S^{2m-1} in $\mathbb{C}^m$ by isometries with isotropy subgroup $\mathrm{U}(m-1)$; $\mathrm{U}(m-1) \to \mathrm{U}(m) \to S^{2m-1}$ is a principle $\mathrm{U}(m-1)$ bundle.

The general linear group $\mathrm{GL}(m+1, \mathbb{R})$ acts naturally and transitively on $\mathbb{R}^{m+1} - \{0\}$ with the isotropy subgroup $\mathrm{GL}(m, \mathbb{R})$; $\mathrm{GL}(m, \mathbb{R}) \to \mathrm{GL}(m+1, \mathbb{R}) \to \mathbb{R}^{m+1} - \{0\}$ is a principle $\mathrm{GL}(m, \mathbb{R})$ bundle.

Complex projective space $\mathbb{CP}^{m-1} = S^{2m-1}/S^1 = \mathrm{U}(m)/\,\mathrm{U}(m-1) \times \mathrm{U}(1)$. We take

$$A = \mathrm{U}(m-1), \quad B = \mathrm{U}(m-1) \times \mathrm{U}(1), \quad C = \mathrm{U}(m)$$

in Theorem 7.3 to obtain the Hopf fibration:

$$\begin{aligned} S^1 &= \mathrm{U}(m-1) \times \mathrm{U}(1)/\,\mathrm{U}(m-1) = B/A \\ \to S^{2m-1} &= \mathrm{U}(m)/\,\mathrm{U}(m-1) = C/A \\ \to \mathbb{CP}^{m-1} &= \mathrm{U}(m)/\,\mathrm{U}(m-1) \times \mathrm{U}(1) = C/B\,. \end{aligned}$$

The Grassmannian $\mathrm{Gr}_k(V)$ is the set of all k-dimensional subspaces of a real or complex vector space V. Let $(\cdot,\cdot)$ be a positive definite inner product on V. If $\sigma \in V$, let $\pi(\sigma)$ denote orthogonal projection on V. This identifies $\mathrm{Gr}_k(V)$ with the set of orthogonal projections of rank k in $\mathrm{Hom}(V)$. The orthogonal group $\mathrm{O}(V, (\cdot,\cdot))$ acts transitively on $\mathrm{Gr}_k(V)$ and the isotropy subgroup is $\mathrm{O}(\sigma) \times \mathrm{O}(\sigma^\perp)$. We have

$$\begin{aligned} \mathrm{Gr}_k(\mathbb{R}^m) &= \mathrm{O}(m)/\{\mathrm{O}(k) \times \mathrm{O}(m-k)\}, \\ \mathrm{Gr}_k(\mathbb{C}^m) &= \mathrm{U}(m)/\{\mathrm{U}(k) \times \mathrm{U}(m-k)\}\,. \end{aligned}$$

We take $k = 1$ and $V = \mathbb{R}^{m+1}$ (resp. $V = \mathbb{C}^{m+1}$) to obtain real (resp. complex) projective space $\mathbb{RP}^m$ (resp. $\mathbb{CP}^m$). The earliest work on these spaces is due to Julius Plücker and the more general framework is due to Hermann Grassmann.

H. Grassmann (1809–1877) J. Plücker (1801–1868)

7.2 THE ISOMETRY GROUP

A diffeomorphism Ψ of a smooth pseudo-Riemannian manifold (M, g) such that $\Psi^*(g) = g$ is said to be an *isometry*. Such maps form a group under composition and we let $\mathcal{I}(M, g)$ be the group of isometries. The following is a useful remark.

Lemma 7.4 Let Ψ_1 and Ψ_2 be isometries of a connected pseudo-Riemannian manifold (M, g). If there exists $P \in M$ so $\Psi_1(P) = \Psi_2(P)$ and $(\Psi_1)_*(P) = (\Psi_2)_*(P)$, then $\Psi_1 = \Psi_2$.

Proof. Set $\Psi := \Psi_1^{-1}\Psi_2$. Then $\Psi(P) = P$ and $\Psi_*(P) = \mathrm{Id}$. Let

$$S = \{Q \in M : \Psi(Q) = Q \text{ and } \Psi_*(Q) = \mathrm{Id}\}\,.$$

Then S is a non-empty closed subset of M. Since $\Psi_*(Q) = \mathrm{Id}$, Ψ preserves the geodesics through Q and hence $\Psi = \mathrm{Id}$ near Q. Consequently, S is open. Since M is connected, $S = M$ so $\Psi = \mathrm{Id}$. □

Lemma 7.5 Let P be a point of a connected Riemannian manifold (M, g). Use $\exp_P^g$ to identify $B_{5r}^g(0)$ in T_PM with a neighborhood of P in M for some $r > 0$.

1. Let g_s for $s \in [-1, 1]$ be a continuous 1-parameter family of Riemannian metrics on M with $g_0 = g$. Then $d_M^{g_s}(P, \cdot)$ converges uniformly to $d_M^g(P, \cdot)$ on $B_r^g(P)$ as $s \to 0$.
2. $\lim_{t\to 0} \|t\xi - t\eta\| \cdot \{d_M^g(t\xi, t\eta)\}^{-1} = 1$ for all $\xi, \eta \in T_PM$.
3. If Ψ is a map from M to M which preserves distance, then $\Psi \in \mathcal{I}(M, g)$.

Proof. Let $0 < \epsilon < \frac{1}{5}$ be given. Since $\bar{B}_{3r}(P)$ is a compact subset of $B_{5r}(P)$, we may choose $\delta > 0$ so $|t| < \delta$ and $|s| < \delta$ implies

$$(1 - \epsilon)g_s < g_t < (1 + \epsilon)g_s \quad \text{on} \quad B_{3r}(P)\,. \tag{7.2.a}$$

Let $Q \in B_r(P)$. Let α_0 be the g_0 geodesic from P to Q. Then

$$d_M^{g_t}(P,Q) \le L_{g_t}(\alpha_0) \le (1+\epsilon)L_{g_0}(\alpha_0) \le \tfrac{6}{5}r\,. \tag{7.2.b}$$

Let α_t be a curve in M from P to Q with $|L_{g_t}(\alpha_t) - d_M^{g_t}(P,Q)| < \epsilon r$. We then have that

$$L_{g_t}(\alpha_t) - \epsilon r \le d_M^{g_t}(P,Q)\,. \tag{7.2.c}$$

Suppose α_t leaves $B_{2r}^{g_0}(P)$. Let β_t be the part of α_t from Q to the boundary of $B_{2r}^{g_0}(P)$. Then

$$\begin{aligned} \tfrac{6}{5}r &\ge d_M^{g_t}(P,Q) \ge L_{g_t}(\alpha_t) - \epsilon r \ge L_{g_t}(\beta_t) - \epsilon r \ge (1-\epsilon)L_{g_0}(\beta_t) - \epsilon r \\ &\ge (1-\epsilon)2r - \epsilon r = (2-3\epsilon)r \ge (2-\tfrac{3}{5})r = \tfrac{7}{5}r \end{aligned}$$

which is false. Therefore, α_t remains in $B_{3r}^{g}(P)$. Equations (7.2.a), (7.2.b), and (7.2.c) yield:

$$\begin{aligned} d_M^{g_t}(P,Q) &\ge L_{g_t}(\alpha_t) - \epsilon r \ge (1-\epsilon)L_{g_s}(\alpha_t) - \epsilon r \ge (1-\epsilon)d_M^{g_s}(P,Q) - \epsilon r \\ &\ge d_M^{g_s}(P,Q) - \tfrac{11}{5}\epsilon r\,. \end{aligned}$$

Interchanging the roles of t and s permits us to establish Assertion 1 by estimating:

$$|d_M^{g_t}(P,Q) - d_M^{g_s}(P,Q)| \le \tfrac{11}{5}\epsilon r\,.$$

We work in normal coordinates. Let $h_{t,ij}(Q) := g_{ij}(tQ)$. Then h_0 is the constant linear inner product given by $g(0)$ on T_PM. Let $\Psi_t(Q) = tQ$ be a dilation; $\Psi_t^*g = t^2h_t$ for $t > 0$. We derive Assertion 2 from Assertion 1 by computing:

$$\begin{aligned} \lim_{t\to 0} \|t\xi - t\eta\| \cdot \{d_M^g(t\xi, t\eta)\}^{-1} &= \lim_{t\to 0} t d_{h_0}^M(\xi,\eta) \cdot \{t d_{h_t}^M(\xi,\eta)\}^{-1} \\ &= \lim_{t\to 0} d_{h_0}^M(\xi,\eta) \cdot \{d_{h_t}^M(\xi,\eta)\}^{-1} = 1\,. \end{aligned}$$

We now prove Assertion 3. Let Ψ be a distance preserving map from M to M; Ψ is continuous since $d(P,Q) < \epsilon$ implies $d(\Psi(P),\Psi(Q)) < \epsilon$. Fix $P \in M$. Let $\tilde{P} := \Psi(P)$ and let $\tilde{Q} := \Psi(Q)$. Choose $\epsilon > 0$ so if $d(P,Q) < \epsilon$ (resp. $d(\tilde{P},\tilde{Q}) < \epsilon$), then there is a unique shortest geodesic γ (resp. $\tilde{\gamma}$) from P to Q (resp. $\tilde{P}$ to $\tilde{Q}$). Let $\delta = d(P,Q) < \epsilon$. Let $R = \gamma(t)$ for $0 < t < \delta$ and let $\tilde{R} := \Psi(R)$. Then

$$d(P,R) = t \quad \text{and} \quad d(Q,R) = \delta - t\,.$$

Applying Ψ yields $d(\tilde{P},\tilde{R}) = t$, $d(\tilde{R},\tilde{Q}) = \delta - t$, and $d(\tilde{P},\tilde{Q}) = \delta$. The properties of the minimizing geodesic for Riemannian geometry which were established in Book I now imply $\tilde{R} = \tilde{\gamma}(t)$. This shows that Ψ maps geodesics to geodesics.

We use the exponential map to identify a neighborhood of 0 in T_PM with a neighborhood of P in M and similarly to identify a neighborhood of 0 in $T_{\tilde{P}}M$ with a neighborhood of $\tilde{P}$ in M and to express $\Psi = \exp_{\tilde{P}} \circ \psi \circ \log_P$ where ψ is continuous map from $(T_PM, 0)$ to $(T_{\tilde{P}}M, 0)$;

the above analysis shows that $\psi(tv) = t\psi(v)$ for any $t \in \mathbb{R}$ and that $\|\psi(v)\| = \|v\|$. We use the polarization identity, Assertion 1, and the fact that ψ is distance preserving to compute:

$$\begin{aligned}\frac{2(\xi,\eta)}{\|\xi\|\ \|\eta\|}(P) &= \frac{\|\xi\|^2+\|\eta\|^2-\|\xi-\eta\|^2}{\|\xi\|\ \|\eta\|}(P) = \frac{\|\xi\|^2+\|\eta\|^2}{\|\xi\|\ \|\eta\|}(P) - \lim_{t\to 0}\frac{d(t\xi,t\eta)^2}{\|t\xi\|\ \|t\eta\|}(P)\\ &= \frac{\|\tilde\xi\|^2+\|\tilde\eta\|^2}{\|\tilde\xi\|\ \|\tilde\eta\|}(\tilde P) - \lim_{t\to 0}\frac{d(t\tilde\xi,t\tilde\eta)^2}{\|t\tilde\xi\|\ \|t\tilde\eta\|}(\tilde P) = \frac{\|\tilde\xi\|^2+\|\tilde\eta\|^2-\|\tilde\xi-\tilde\eta\|^2}{\|\tilde\xi\|\ \|\tilde\eta\|}(\tilde P) = \frac{2(\tilde\xi,\tilde\eta)}{\|\tilde\xi\|\ \|\tilde\eta\|}(\tilde P).\end{aligned}$$

This shows that $(\xi,\eta) = (\tilde\xi,\tilde\eta)$ as well. But $\xi+\eta$ is determined by $\{\|\xi\|, \|\eta\|, (\xi,\eta)\}$. This shows $\psi(\xi+\eta) = \psi(\xi)+\psi(\eta)$. Consequently, ψ is an orthogonal transformation and Ψ is a smooth local isometry. Since Ψ preserves distances, Ψ is injective. Since Ψ is a local isometry, range$\{\Psi\}$ is open. Since Ψ is distance preserving, range$\{\Psi\}$ is closed. Since range$\{\Psi\}$ is non-empty and M is connected, range$\{\Psi\} = M$. It now follows Ψ is an isometry. □

Theorem 7.6

1. *If (M,g) is a pseudo-Riemannian manifold, then $\mathcal{I}(M,g)$ is a Lie group and the natural action of $\mathcal{I}(M,g)$ on M is smooth. A group homomorphism f from $\mathbb{R}$ to $\mathcal{I}(M,g)$ is smooth if and only if the map $(t,x)\to f(t)x$ from $\mathbb{R}\times M$ to M is smooth.*
2. *Every compact Lie group can be realized as the full group of isometries of a compact Riemannian manifold.*
3. *If (M,g) is a compact Riemannian manifold, then $\mathcal{I}(M,g)$ is compact.*

Proof. We refer to Myers and Steenrod [45] and to Palais [50] for the proof of Assertion 1, and to Saerens and Zame [55] for the proof of Assertion 2 as these results are beyond the scope of this book. We establish Assertion 3 as follows. If (M,g) is a compact Riemannian manifold, then $\mathcal{I}(M,g)$ has the metric topology given by

$$d_M^g(\Psi_1,\Psi_2) := \max_{P\in M} d_M^g(\Psi_1 P, \Psi_2 P).$$

Find a countable dense subset $\{S_1, S_2, \dots\}$ of M. Let Ψ_n be a sequence of isometries. By passing to a subsequence, we may assume $\lim_{n\to\infty}\Psi_n S_i$ exists for every i. Let $\epsilon > 0$ be given. As M is compact, there exists $n = n(\epsilon)$ so

$$M = B_\epsilon(S_1)\cup\dots\cup B_\epsilon(S_n).$$

Choose $N = N(\epsilon)$ so that if $a, b > N$, then $d(\Psi_a S_i, \Psi_b S_i) < \epsilon$ for $1 \le i \le n$. Given $P \in M$, choose i with $1 \le i \le n$ so $d(P, S_i) < \epsilon$. If $a, b > N$, we may estimate:

$$\begin{aligned} d(\Psi_a P, \Psi_b P) &\le d(\Psi_a P, \Psi_a S_i) + d(\Psi_a S_i, \Psi_b S_i) + d(\Psi_b S_i, \Psi_b P)\\ &\le d(P, S_i) + \epsilon + d(S_i, P) \le 3\epsilon.\end{aligned}$$

This shows $d(\Psi_a, \Psi_b) < 3\epsilon$ if $a, b > N$. Consequently, the sequence $\{\Psi_n\}$ is uniformly Cauchy and converges uniformly to a continuous distance preserving map Ψ from M to M. Lemma 7.5 implies $\Psi \in \mathcal{I}(M,g)$. This shows $\mathcal{I}(M,g)$ is compact. □

7.2.1 EXAMPLE. Assertion 3 of Theorem 7.6 fails in the higher signature setting. Let (x^1, x^2, y^1, y^2) be the usual periodic parameters on the 4-dimensional torus $\mathbb{T}^4 := \mathbb{R}^4/\mathbb{Z}^4$. Let $ds^2 = dx^1 \circ dy^1 + dx^2 \circ dy^2$ be a flat metric of signature $(2, 2)$. The torus is a compact Abelian Lie group under addition and ds^2 is a bi-invariant metric on $\mathbb{T}^4$. There is a canonical action of $\mathrm{SL}(2, \mathbb{Z})$ on $\mathbb{T}^4$. If $A \in \mathrm{SL}(2, \mathbb{Z})$, we set $\sigma(A) = A \oplus (A^{-1})^t$ to define a map from $\mathrm{SL}(2, \mathbb{Z})$ to $\mathrm{SL}(4, \mathbb{Z})$ and thereby define an action of $\mathrm{SL}(2, \mathbb{Z})$ on $\mathbb{T}^4$. If $A = (n_{ij})$, then

$$A \cdot (x^1, x^2, y^1, y^2) := (n_{11}x^1 + n_{12}x^2, n_{21}x^1 + n_{22}x^2, n_{22}y^1 - n_{21}y^2, -n_{12}y^1 + n_{11}y^2)\,.$$

It is clear that A acts by isometries on $\mathbb{T}^4$. The fundamental group of $\mathbb{T}^4$ is $\mathbb{Z}^4$ and the induced action of A on the fundamental group is given by σ. Consequently, all the isometries $\sigma(A)$ belong to different arc components of $\mathcal{I}(\mathbb{T}^4, ds^2)$ so $\mathcal{I}(\mathbb{T}^4, ds^2)$ has an infinite number of arc components. We refer to work of Melnick [42] to see that there exist compact Lorentzian manifolds such that the connected component of the isometry group is non-compact.

7.2.2 EXAMPLE. Give the upper-half plane $\mathbb{H}^2 := \{(x, y) \in \mathbb{R}^2 : y > 0\}$ the hyperbolic metric $ds^2 := y^{-2}(dz \circ d\bar{z})$. Let

$$\mathrm{PSL}(2, \mathbb{R}) := \mathrm{SL}(2, \mathbb{R})/\mathbb{Z}_2 := \{\pm 1\}$$

be the *projective special linear group*. The *linear fractional transformations*

$$T_A(z) := \frac{az + b}{cz + d} \quad \text{for} \quad A = \begin{pmatrix} a & b \\ c & d \end{pmatrix} \in \mathrm{SL}(2, \mathbb{R})$$

define an action of $\mathrm{PSL}(2, \mathbb{R})$ on $\mathbb{H}$ by orientation preserving isometries and every orientation preserving isometry of $\mathbb{H}^2$ arises in this way. If we identify $\mathbb{H}^2$ with the unit pseudo-sphere in Minkowski space, then we identify $\mathrm{PSL}(2, \mathbb{R})$ with the connected component of the identity in $\mathrm{O}(1, 2)$. $\mathrm{PSL}(2, \mathbb{R})$ acts transitively on the unit tangent bundle of the hyperbolic plane by isometries and we may identify $\mathrm{PSL}(2, \mathbb{R})$ with the unit sphere bundle of the tangent bundle to $\mathbb{H}^2$.

We showed in Lemma 6.25 that the Killing form is a bi-invariant Lorentzian metric on $\mathrm{PSL}(2, \mathbb{R})$. Let Σ be an orientable Riemann surface with a hyperbolic metric of constant Gaussian curvature -1. The universal cover of Σ is then $\mathbb{H}^2$ and the deck group Γ is a cocompact lattice Γ in $\mathrm{PSL}(2, \mathbb{R})$. We then have that $M := \mathrm{PSL}(2, \mathbb{R})/\Gamma$ is a compact manifold which admits a Lorentzian metric which is invariant under the left action of $\mathrm{PSL}(2, \mathbb{R})$. Therefore, $\mathrm{PSL}(2, \mathbb{R})$ acts by isometries on (M, g) and the natural map from $\mathrm{PSL}(2, \mathbb{R})$ to $\mathcal{I}(M, g)$ is smooth and non-trivial. If $\mathcal{I}(M, g)$ is a compact Lie group, then by the Peter–Weyl Theorem, the matrix coefficients of the unitary representations of $\mathcal{I}(M, g)$ separate points of $\mathcal{I}(M, g)$. This would imply that there exists a non-trivial orthogonal representation of $\mathrm{PSL}(2, \mathbb{R})$ which contradicts Lemma 6.25.

7.3 THE LIE DERIVATIVE AND KILLING VECTOR FIELDS

We present concepts originally introduced by the German mathematician Wilhelm Karl Joseph Killing and by the Norwegian mathematician Sophus Lie.

W. Killing (1847–1923)

S. Lie (1842–1899)

7.3.1 THE LIE DERIVATIVE. Let Φ_t^X be the flow for a vector field X on M; it is only locally defined but this plays no role. Let $S = Y \otimes \xi \in C^\infty((\otimes^k TM) \otimes (\otimes^\ell T^*M))$ be a mixed tensor field where Y is a smooth section to $\otimes^k TM$ and ξ is a smooth section to $\otimes^\ell T^*M$. The *Lie derivative* $\mathcal{L}_X$ of $Y \otimes \xi$ is defined by setting:

$$\begin{aligned}&\mathcal{L}_X(Y \otimes \xi) := \mathcal{L}_X(Y) \otimes \xi + Y \otimes \mathcal{L}_X(\xi) \quad \text{where}\\ &\mathcal{L}_X(Y) := \lim_{t\to 0} t^{-1}\{(\Phi_{-t}^X)_* Y - Y\} \quad \text{and} \quad \mathcal{L}_X(\xi) := \lim_{t\to 0} t^{-1}\{(\Phi_t^X)^*\xi - \xi\}.\end{aligned}$$

It is then immediate that $\mathcal{L}_X$ satisfies the *Leibnitz formula*:

$$\mathcal{L}_X(S_1 \otimes S_2) = \mathcal{L}_X(S_1) \otimes S_2 + S_1 \otimes \mathcal{L}_X(S_2) \tag{7.3.a}$$

for any tensor fields S_1 and S_2. If $\alpha_p \in C^\infty(\Lambda^p T^*M)$ and if $\beta_q \in C^\infty(\Lambda^q T^*M)$, then:

$$\mathcal{L}_X(\alpha_p \wedge \beta_q) = \mathcal{L}_X(\alpha_p) \wedge \beta_q + \alpha_p \wedge \mathcal{L}_X(\beta_q) \quad \text{and} \quad d\mathcal{L}_X(\alpha_p) = \mathcal{L}_X(d\alpha_p).$$

If $f \in C^\infty(M)$, then $\mathcal{L}_X(f) = X(f)$. We now establish some results for the Lie derivative; there are many other identities but they can all be derived easily using the same techniques as we shall use to prove Lemma 7.7.

Lemma 7.7 Let $X \in C^\infty(TM)$, $Y \in C^\infty(TM)$, and $\omega \in C^\infty(T^*M)$. Then

1. $X\langle Y, \omega\rangle = \langle \mathcal{L}_X Y, \omega\rangle + \langle Y, \mathcal{L}_X \omega\rangle$. 2. $\mathcal{L}_X Y = [X, Y]$. 3. $[\mathcal{L}_X, \mathcal{L}_Y] = \mathcal{L}_{[X,Y]}$.

Proof. Suppose first that $X(P) \neq 0$. Choose local coordinates $\vec{x} = (x^1, \dots, x^m)$ near P so X corresponds to ∂_{x^1}; the flow then takes the form

$$\Phi_t^X(\vec{x}) = (x^1 + t, x^2, \dots, x^m). \tag{7.3.b}$$

Let $Y = b^j \partial_{x^j}$ and let $\omega = c_k dx^k$. Then:

$$\begin{aligned}&\{(\Phi_{-t}^X)_* Y\}(\vec{x}) = b^j(x^1 + t, x^2, \dots, x^m)\partial_{x^j}, \quad \mathcal{L}_X(Y) = \{\partial_{x^1} b^j\}\partial_{x^j},\\ &\{\Phi_t^* \omega\}(\vec{x}) = c_k(x^1 + t, x^2, \dots, x^m) dx^k, \qquad \mathcal{L}_X(\omega) = \{\partial_{x^1} c_k\} dx^k.\end{aligned}$$

Assertion 1 and Assertion 2 now follow at P. More generally, suppose $X = a^i \partial_{x^i}$ where $a(P)$ is zero. We use Lemma 6.8 to see

$$\Phi_t^X(\vec{x}) = \vec{x} + ta(\vec{x}) + \tfrac{1}{2}t^2 da(\vec{x}) \cdot a(\vec{x}) + O(t^3)\,.$$

This shows that $\mathcal{L}_X(Y)(P)$ and $\mathcal{L}_X(\omega)(P)$ depend smoothly on the 1-jets of Y at P, the 1-jets of ω at P, and the 1-jets of X at P. If X vanishes identically near P, then $\Phi_t^X(Q) = Q$ for Q near P and $\mathcal{L}_X(Y)$ and $\mathcal{L}_X(\omega)$ vanish identically near P; Assertion 1 and Assertion 2 then hold trivially. If there exists a sequence of points $P_n \to P$ with $X(P_n) \neq 0$, then Assertion 1 and Assertion 2 hold at P_n and hence by continuity at P. This establishes Assertion 1 and Assertion 2 in generality.

If Z is a smooth vector field, then Assertion 2 and the Jacobi identity imply that:

$$[\mathcal{L}_X, \mathcal{L}_Y]Z = [X,[Y,Z]] - [Y,[X,Z]] = [[X,Y],Z] = \mathcal{L}_{[X,Y]}(Z)\,.$$

This proves Assertion 3 for vector fields; Assertion 3 for 1-forms then follows by duality from Assertion 1 and for general mixed tensor fields from Equation (7.3.a). □

7.3.2 KILLING VECTOR FIELDS.

Lemma 7.8 Let (M, g) be a connected pseudo-Riemannian manifold. If $X \in C^\infty(TM)$, let $A_X(Y) := -\nabla_Y X$.

1. The following conditions are equivalent and if any is satisfied, then X is said to be a *Killing vector field*:
   (a) $\mathcal{L}_X(g) = 0$.
   (b) The local flows Φ_t^X defined by X preserve the metric g. Thus, these local flows are isometries.
   (c) A_X is skew-symmetric.
2. If X is a Killing vector field, then:
   (a) $g(X(\sigma(s)), \dot\sigma(s))$ is constant for any geodesic σ.
   (b) If there exists $P \in M$ so $X(P) = 0$ and so $\nabla X(P) = 0$, then $X \equiv 0$.
3. If X and Y are Killing vector fields, then $[X, Y]$ is a Killing vector field.

Proof. If the local flows Φ_t^X preserve g, then $\mathcal{L}_X(g) = 0$. Conversely, suppose that $\mathcal{L}_X(g) = 0$. If $X(P) \neq 0$, then we may choose local coordinates $\vec{x} = (x^1, \dots, x^m)$ so $X = \partial_{x^1}$. The condition that $\mathcal{L}_X(g) = 0$ then implies $g_{ij}(x^1, \dots, x^m) = g_{ij}(x^2, \dots, x^m)$ so the flows given in Equation (7.3.b) are isometries. The same continuity argument used to prove Lemma 7.7 can then be used to deduce this result even if $X(P) = 0$. This proves the equivalence of Assertion 1-a and Assertion 1-b. We prove the equivalence of Assertion 1-a and Assertion 1-c by using Lemma 7.7 to compute:

$$\begin{aligned}(\mathcal{L}_X g)(Y,Z) &= (\mathcal{L}_X g)(Y\otimes Z) = X\cdot g(Y\otimes Z) - g(\mathcal{L}_X(Y\otimes Z))\\ &= X\cdot g(Y,Z) - g([X,Y],Z) - g([X,Z],Y)\\ &= X\cdot g(Y,Z) - g(\nabla_X Y,Z) + g(\nabla_Y X,Z) - g(\nabla_X Z,Y) + g(\nabla_Z Y,X)\\ &= (\nabla_X g)(Y,Z) - g(A_X Y,Z) - g(A_X Z,Y)\\ &= -\{g(A_X Y,Z) + g(A_X Z,Y)\}\,.\end{aligned}$$

Let X be a Killing vector field and let σ be a geodesic. By Assertion 1-c, A_X is skew-symmetric so $g(\dot\sigma, A_X\dot\sigma) = 0$. We establish Assertion 2-a by computing:

$$\partial_s g(\dot\sigma, X) = g(\nabla_{\dot\sigma}\dot\sigma, X) + g(\dot\sigma, \nabla_{\dot\sigma} X) = 0 - g(\dot\sigma, A_X\dot\sigma) = 0\,.$$

Assertion 2-b is the analog of Lemma 7.4 on the infinitesimal level. If $X(P) = 0$, then the map $(\Phi_t^X)_*$ from $\mathbb{R}$ to $\mathrm{GL}(T_P M)$ is a group homomorphism and the corresponding representation on the Lie algebra level is given by $\partial_{t=0}(\Phi_t^X)_*$ mapping the Lie algebra $\mathfrak{r}$ of $\mathbb{R}$ to the Lie algebra $\mathfrak{gl}$ of GL. Choose local coordinates $\vec{x} = (x^1,\dots,x^m)$. By Lemma 6.8, $\Phi_t^X(\vec{x}) = \vec{x} + ta(\vec{x}) + O(t^2)$. Therefore,

$$\begin{aligned}(\Phi_t^X)_*(P)(v) &= \lim_{\epsilon\to 0}\epsilon^{-1}\{\Phi_t^X(P+\epsilon v) - \Phi_t^X(P)\}\\ &= v + t\lim_{\epsilon\to 0}\epsilon^{-1}\{a(P+\epsilon v) - a(P)\} + O(t^2)\,.\end{aligned}$$

Since $\nabla X(P) = 0$, $da(P) = 0$ so $(\Phi_t^X)_*(P)(v) = \mathrm{Id} + O(t^2)$. This implies that

$$\partial_{t=0}\{\Phi_t^X\}_*(P) = 0\,.$$

Consequently, $\partial_{t=0}(\Phi_t^X)_*$ is the trivial map and $\{\Phi_t^X\}(P)_* = \mathrm{Id}$ for all t. As the maps Φ_t^X are local isometries, they commute with $\exp_P$, i.e.,

$$\Phi_t^X \circ \exp_P = \exp_P \circ (\Phi_t^X)_* = \exp_P\,.$$

Therefore, $\Phi_t = \mathrm{Id}$ near P so $X = 0$ near P. This proves Assertion 2-b. By Lemma 7.7, we have that $[\mathcal{L}_X, \mathcal{L}_Y] = \mathcal{L}_{[X,Y]}$. Since $\mathcal{L}_X(g) = 0$ and $\mathcal{L}_Y(g) = 0$, $\mathcal{L}_{[X,Y]}g = 0$. Assertion 3 follows. □

7.3.3 THE LIE ALGEBRA OF A PSEUDO-RIEMANNIAN MANIFOLD. Denote the set of Killing vector fields by $\mathcal{K} = \mathcal{K}(M,g)$. By Lemma 7.8, $\mathcal{K}$ is a real vector space which is closed under bracket and forms a Lie algebra. Let Θ_t be a smooth 1-parameter family of isometries in $\mathcal{I}(M,g)$; Θ_t is the 1-parameter family generated by a Killing vector field. Therefore, the Lie algebra of the group of isometries of $\mathcal{I}(M,g)$ is a Lie subalgebra of $\mathcal{K}$. However, every element of $\mathcal{K}$ need not correspond to a 1-parameter family of isometries. For example, if $(M,g) = ((0,1), dx^2)$, then $X = \partial_x$ is a Killing vector field but $\mathcal{I}(M,g) = \mathbb{Z}_2$ is 0-dimensional.

Lemma 7.9 Let (M,g) be a pseudo-Riemannian manifold of dimension m. If M is compact or if (M,g) is geodesically complete, then every Killing vector field defines a 1-parameter subgroup of isometries of M and $\mathcal{K}$ is the Lie algebra of the isometry group $\mathcal{I}(M,g)$.

Proof. Let $X \in \mathcal{K}$. If M is compact, then the flow Φ_t for X on M is globally defined for all t and defines a 1-parameter flow by isometries; this proves the Lemma in this case. Next, suppose (M, g) is complete. We consider the forward flow $t \geq 0$ as the case $t \leq 0$ can be handled by replacing X by $-X$. For each point $P \in M$, let $\epsilon(P) > 0$ be chosen maximal so that Φ_t is defined for $0 \leq t < \epsilon(P)$. The maximal domain $\mathcal{O}$ for Φ is

$$\mathcal{O} := \cup_{P \in M}\{[0, \epsilon(P)) \times P\}.$$

As $\liminf_{Q \to P} \epsilon(Q) = \epsilon(P)$, $\mathcal{O}$ is open; Φ is smooth on $\mathcal{O}$ and $\Phi_s \Phi_t = \Phi_{s+t}$ where defined. We must show $\epsilon(P) = \infty$ for all P. Suppose to the contrary that $\epsilon = \epsilon(P) < \infty$ for some P and argue for a contradiction. The exponential map $\exp_P$ is a diffeomorphism from a neighborhood of 0 in $T_P M$ to a neighborhood of P in M. Let $\log_P$ be the local inverse. Choose $0 < \delta \leq \frac{1}{2}\epsilon$ so that $\Phi_t(P) \in \text{domain}\{\log_P\}$ for $0 \leq t \leq \delta$. Let $\epsilon - \delta \leq t < \epsilon$. Since the maps Φ_t are local isometries, $\Phi_t \circ \exp_P = \exp_{\Phi_t P} \circ (\Phi_t(P))_*$. As (M, ∇) is complete, we may express:

$$\begin{aligned} \Phi_t(P) &= \Phi_{\epsilon-\delta}\Phi_{t+\delta-\epsilon}P = \Phi_{\epsilon-\delta}\exp_P\{\log_P \Phi_{t+\delta-\epsilon}P\} \\ &= \exp_{\Phi_{\epsilon-\delta}P}\{(\Phi_{\epsilon-\delta}(P))_* \log_P \Phi_{t+\delta-\epsilon}P\}. \end{aligned}$$

The right-hand side of the equation is well-defined and smooth for $\epsilon - \delta \leq t \leq \epsilon$. This shows that $[0, \epsilon)$ was not a maximal domain for $\Phi_t(P)$. □

Let $\mathcal{I}_0(M, g)$ be the connected component of the identity in the isometry group $\mathcal{I}(M, g)$. We say that a connected pseudo-Riemannian manifold (M, g) is *homogeneous* if $\mathcal{I}_0(M, g)$ acts transitively on M. We say that (M, g) is *locally homogeneous* if given any pair of points P and Q in M, there exists an isometry from some neighborhood of P to some neighborhood of Q in M. Observe that if G is a connected Lie group which is equipped with a left-invariant non-degenerate bilinear form, then the left action of G is a transitive action by isometries and, consequently, (G, g) is homogeneous.

Lemma 7.10 Let (M, g) be a pseudo-Riemannian manifold of dimension m.

1. If (M, g) is homogeneous, then there are m linearly independent Killing vector fields at every point of M.

2. Assume that M is compact or that (M, g) is complete. If there are m linearly independent Killing vector fields at every point of M, then (M, g) is homogeneous.

Proof. Assume (M, g) is homogeneous so $\mathcal{I}_0(M, g)$ acts transitively on M. By Theorem 7.3, the map $\mathcal{I}_0(M, g) \to \mathcal{I}_0(M, g) \cdot P$ is a submersion so the natural map from the Lie algebra of $\mathcal{I}_0(M, g)$ to $T_P M$ is surjective for any $P \in M$. Consequently, there exist m linearly independent Killing vector fields at P. Conversely, assume that there exist m linearly independent Killing vector fields at every point P of M. We apply Lemma 7.9 to see every Killing vector field defines a smooth 1-parameter family of isometries. It now follows that $\mathcal{I}_0(M, g) \cdot P$ contains an open neighborhood of P. Since M is connected, it follows $\mathcal{I}_0(M, g)$ acts transitively on M and (M, g) is homogeneous. □

The hypothesis that M is either compact or complete is necessary to ensure that the conclusions of either Lemma 7.9 or Lemma 7.10 hold. The vector field ∂_x on $((0, 1), dx^2)$ is a globally defined Killing vector field but there is no corresponding 1-parameter family of global isometries of $(0, 1)$ and $(0, 1)$ is not globally homogeneous.

Results of Kobayashi [37] and of Nomizu [48] provide a useful characterization of homogeneity (see also Hall [27] in the Lorentzian setting). We first introduce a bit of notation. Let $W = TM \oplus (T^*M \otimes TM)$; this is a vector bundle of dimension $m + m^2$. Let X be a Killing vector field defined on an open subset $\mathcal{O}$ of M; X will be called a *local Killing vector field*. If $P \in \mathcal{O}$, set

$$\mathfrak{S}_X(P) := X(P) \oplus \nabla X(P) \in W_P .$$

Let $\mathfrak{K}_P = \cup_X \mathfrak{S}_X(P)$. Since the sum of two Killing vector fields or a constant times a Killing vector field is again a Killing vector field, $\mathfrak{K}_P$ is a linear subspace of W. Let $\mathfrak{K} := \cup_P \mathfrak{K}_P$. If (M, g) is homogeneous, then $\dim\{\mathfrak{K}(\cdot)\}$ is constant. The following result provides a partial converse; it is necessary to assume M simply connected to avoid problems with the fundamental group.

Theorem 7.11 *Let (M, g) be a simply connected pseudo-Riemannian manifold. Assume* $\dim\{\mathfrak{K}(\cdot)\}$ *is constant on M. Then $\mathfrak{K}$ is a trivial vector bundle over M and every local Killing vector field on M extends to a global Killing vector field on M.*

Proof. Let $\nu = \dim\{\mathfrak{K}(\cdot)\}$. Let P be a point of M. Choose local Killing vector fields $\{X_1, \ldots, X_\nu\}$ so that $\{\mathfrak{S}_{X_1}(P), \ldots, \mathfrak{S}_{X_\nu}(P)\}$ forms a basis for $\mathfrak{K}_P$. By continuity, $\{\mathfrak{S}_{X_1}(Q), \ldots, \mathfrak{S}_{X_\nu}(Q)\}$ are linearly independent elements of W_Q for Q in some neighborhood $\mathcal{O}$ of P. Since $\dim\{\mathfrak{K}_Q\} = \nu$, these elements span $\mathfrak{K}_Q$ and, therefore, $\{\mathfrak{S}_{X_1}, \ldots, \mathfrak{S}_{X_\nu}\}$ is a frame for $\mathfrak{K}(\mathcal{O})$. This shows that $\mathfrak{K}(M)$ is a smooth vector subbundle of W.

Let $\{Y_1, \ldots, Y_\nu\}$ be another collection of local Killing vector fields such that

$$\{\mathfrak{S}_{Y_1}(P), \ldots, \mathfrak{S}_{Y_\nu}(P)\}$$

is a basis for $\mathfrak{K}_P$. Express $\mathfrak{S}_{Y_i}(P) = a_i^j \mathfrak{S}_{X_j}(P)$. By Lemma 7.8, a Killing vector field is determined by its value at a point and the value of its covariant derivative at that point. Thus, $Y_i = a_i^j X_j$ and, consequently, $\mathfrak{S}_{Y_i} = a_i^j \mathfrak{S}_{X_j}$ near P. This means that $\mathfrak{K}$ is *locally flat*, i.e., the transition functions of the vector bundle $\mathfrak{K}$ with respect to this canonically defined system of local frame fields are locally constant. Any locally flat vector bundle over a simply connected space is trivial and the trivialization reflects the locally flat structure. In other words, we may use the locally flat structure to extend the Killing vector fields along curves; as M is simply connected, there is no holonomy and this gives a global extension. □

The hypotheses of Theorem 7.11 are clearly necessary. In general, $\dim\{\mathfrak{K}_P\}$ can vary with the point P. For example, let the torus $\mathbb{T}^2$ be given the flat Riemannian metric outside a neighborhood of 0 and have a small radially symmetric bump at the origin. If the bump is chosen properly, then $\dim\{\mathfrak{K}(0)\} = 1$ and $\dim\{\mathfrak{K}_P\} = 3$ for P away from the origin. Therefore, there are local Killing vector fields in this instance which do not extend to global Killing vector fields. The hypothesis that M is simply connected is needed as well. Give $\mathbb{T}^2$ the flat Riemannian metric. The rotation of $\mathbb{R}^2$ about 0 generates a Killing vector field which restricts to a local Killing vector field on $\mathbb{T}^2$ which does not extend to a global Killing vector field.

7.4 HOMOGENEOUS PSEUDO-RIEMANNIAN MANIFOLDS

Lemma 7.12 If G is a compact Lie group which acts on a manifold M, then M admits a G-invariant Riemannian metric. If the action is transitive, then M is compact.

Proof. Let θ be an arbitrary Riemannian metric on M. By Lemma 6.12, we can find a bi-invariant measure $|\operatorname{dvol}|$ on G of total mass 1. If $X, Y \in T_P M$, let

$$\Theta(x, y) := \int_G \theta(g_* x, g_* y)|\operatorname{dvol}|(g)\,.$$

This provides the required invariant Riemannian metric on M. If the action is transitive, then the map $g \to g \cdot P$ is surjective and M is compact. □

Lemma 7.13 Every homogeneous Riemannian manifold is geodesically complete.

Proof. Let G be a Lie group which acts transitively and by isometries on a Riemannian manifold (M, g). Let $P \in M$. Choose $\epsilon > 0$ so that every unit speed geodesic starting at P extends for time at least ϵ. Let $\gamma : [0, T] \to M$ be a unit speed geodesic with $\gamma(0) = P$. Choose $g \in G$ so $L_g P = \gamma(T)$. Since L_g is an isometry, every unit speed geodesic from gP extends for time ϵ. Therefore, we may extend γ to time $[0, T + \epsilon]$ and hence, by induction, to time $[0, \infty)$. □

7.4.1 HOMOGENEOUS PSEUDO-RIEMANNIAN MANIFOLDS. Since compact metric spaces are complete metric spaces, the Hopf–Rinow Theorem (Theorem 3.15 in Book I) shows that every compact Riemannian manifold is geodesically complete. In Section 7.4.2, we will exhibit a Lorentzian metric on the 2-torus that is not geodesically complete, and, in Section 7.4.3, we will exhibit an incomplete homogeneous Lorentzian metric. Consequently, a bit of care must be taken. There is, however, an analog of Lemma 7.13 in the pseudo-Riemannian setting due to Marsden [39].

Theorem 7.14 *Let (M, g) be a compact homogeneous pseudo-Riemannian manifold. Then (M, g) is geodesically complete.*

Proof. Let σ be a geodesic defined on a maximal domain $[0, T)$. We wish to show that $T = \infty$. Suppose to the contrary that $T < \infty$. Let $t_n \to T$ and let $Q_n := \sigma(t_n)$. Since M is compact, by passing to a subsequence, we may assume that $Q_n \to Q$ for some point Q of M. By Lemma 7.10, there exist Killing vector fields $\{X_1, \dots, X_m\}$ so $\{X_1(Q), \dots, X_m(Q)\}$ form a basis for $T_Q M$ and hence form a frame for TM near Q. Expand $\dot\sigma(s) = a^i(s) X_i(\sigma(s))$. The matrix $\xi_{ij} := g(X_i, X_j)$ is a non-singular symmetric matrix near Q. By Lemma 7.8, $g(\dot\sigma(s), X_i(\sigma(s)))$ is independent of s. This implies $g_{ij}(\sigma(s))a^i(s) = c_j$ or equivalently $a^i(s) = g^{ij}(\sigma(s))c_j$. Since g^{ij} is continuous near Q, g^{ij} and hence $a^i(s)$ is bounded near Q. Introduce an auxiliary Riemannian inner product $(\cdot,\cdot)_e$ near Q. We then have $(\dot\sigma, \dot\sigma)_e$ is uniformly bounded near Q. Since the $t_n \to T$ and $Q_n \to Q$, this means that in fact $\sigma(s)$ stays near Q for $t_n < s < T$. Consequently, $\sigma(s)$ tends to Q as $s \to T$. This implies that the maximal domain is not $[0, T)$ and completes the proof. □

7.4.2 AN INCOMPLETE LORENTZIAN METRIC ON THE TORUS.

Every compact Riemannian manifold is geodesically complete. This can fail in the higher signature setting. The following example is due to Meneghini [43]. Let $(N, g_N) := (\mathbb{R}^2 - \{0\}, g_N)$ where

$$g_N := (x^2 + y^2)^{-1} dx \circ dy$$

has signature $(1, 1)$. Let $m(x, y) = (2x, 2y)$. We then have

$$m^* g_N = ((m^*x)^2 + (m^*y)^2)^{-1} d(m^*x) \circ d(m^*y) = (4x^2 + 4y^2)^{-1} 4dx \circ dy = g_N\,.$$

This shows that m is an isometry which generates a cyclic subgroup which acts properly and discontinuously on N. Let $M := N/\mathbb{Z}$ be the quotient by this subgroup; M is a compact manifold which is diffeomorphic to the torus $S^1 \times S^1$ and g_N descends to define a Lorentzian metric g_M on M. One uses the Koszul formula (see Theorem 3.7 and Equation (3.3.a) in Book I) to compute:

$$g(\nabla_{\partial_x}\partial_x, \partial_x) = 0, \quad g(\nabla_{\partial_x}\partial_x, \partial_y) = -\frac{2x}{(x^2 + y^2)^2}, \quad \nabla_{\partial_x}\partial_x = -\frac{2x}{x^2 + y^2}\partial_x\,.$$

Let $\gamma(t) = (\frac{1}{1-t}, 0)$. We compute:

$$\nabla_{\dot\gamma}\dot\gamma = \{\ddot x - 2x^{-1}\dot x \dot x\}\partial_x = \{2(1-t)^{-3} - 2(1-t)(1-t)^{-4}\}\partial_x = 0\,.$$

Therefore, γ is a geodesic with maximal domain $(-\infty, 1)$. Consequently, (N, g_N) is geodesically incomplete so (M, g_M) is geodesically incomplete as well.

7.4.3 AN INCOMPLETE HOMOGENEOUS LORENTZIAN METRIC.

The following example is motivated by work of Guediri [24] and of Guediri and Lafontaine [25]. Let $\mathbb{M}$ be

Minkowski space; $\mathbb{M} := (\mathbb{R}^2, dx \circ dy)$. Let G be the $ax+b$ group of Lemma 6.23. Define a smooth map T from G to $I(\mathbb{M})$ by setting

$$T_{(\alpha,\beta)}(x,y) := (\alpha^{-1}x, \alpha y + \beta)\,.$$

We verify that T is a group homomorphism by computing:

$$\begin{aligned} T_{(\alpha,\beta)} \circ T_{(\gamma,\delta)} : (x,y) \to T_{(\alpha,\beta)}(\gamma^{-1}x, \gamma y + \delta) &= (\alpha^{-1}\gamma^{-1}x, \alpha\gamma y + \alpha\delta + \beta) \\ &= T_{(\alpha\gamma,\alpha\delta+\beta)}(x,y) = T_{(\alpha,\beta)*(\gamma,\delta)}(x,y)\,. \end{aligned}$$

Let $M := \mathbb{R}^+ \times \mathbb{R}$ and let $\mathcal{M} := (M, dx \circ dy)$. Then G acts transitively on $\mathcal{M}$ so $\mathcal{M}$ is a homogeneous Lorentzian manifold. The geodesics of this geometry are straight lines. Through any point of M there is exactly one straight line which is totally contained in M; consequently, M is a geodesically incomplete. Since the map $g \to g \cdot (1,0)$ is G-equivariant diffeomorphism from G to M, $T^*(dx \circ dy)$ is a geodesically incomplete left-invariant Lorentzian metric on the $ax+b$ group.

7.5 LOCAL SYMMETRIC SPACES

There is both a local and a global aspect to the theory of symmetric spaces in the pseudo-Riemannian context. We begin by presenting results of Cartan [9, 10]; the proof of these results uses Jacobi vector fields to examine the local geometry. Let $\mathcal{R}$ be the curvature operator of a pseudo-Riemannian manifold (M,g). We define the *covariant derivative of the curvature operator* $\nabla_X\mathcal{R}$ and the covariant derivative of the curvature tensor ∇R by setting:

$$\begin{aligned} \{(\nabla_X\mathcal{R})(Y,Z)\}W &:= \nabla_X\{\mathcal{R}(Y,Z)W\} - \mathcal{R}(\nabla_X Y, Z)W - \mathcal{R}(Y, \nabla_X Z)W \\ &\quad -\mathcal{R}(Y,Z)\nabla_X W, \\ \nabla R(Y,Z,W,U;X) &:= g(\nabla_X\mathcal{R}(Y,Z)W, U) \\ &= X(R(Y,Z,W,U)) - R(\nabla_X Y, Z, W, U) - R(Y, \nabla_X Z, W, U) \\ &\quad -R(Y,Z,\nabla_X W, U) - R(Y,Z,W,\nabla_X U)\,. \end{aligned}$$

Both ∇R and $\nabla\mathcal{R}$ are tensors. If (M,g) has constant sectional curvature, then $\nabla\mathcal{R} = 0$ as we shall see presently in Example 7.6.1. Consequently, the following result can be regarded as a generalization of Lemma 3.19 in Book I.

Lemma 7.15 Let M and $\tilde{M}$ be pseudo-Riemannian manifolds with curvature operators $\mathcal{R}$ and $\tilde{\mathcal{R}}$, respectively, with $\nabla\mathcal{R} = 0$ and $\tilde{\nabla}\tilde{\mathcal{R}} = 0$. Let $P \in M$ and let $\tilde{P} \in \tilde{M}$. Suppose there exists a linear isometry ψ from $T_P M$ to $T_{\tilde{P}}\tilde{M}$ so that $\psi^*\tilde{\mathcal{R}}(\tilde{P}) = \mathcal{R}(P)$. Then the map

$$\Psi := \exp_{\tilde{P}} \circ \psi \circ \log_P$$

is a local isometry from a neighborhood of P in M to a neighborhood of $\tilde{P}$ in $\tilde{M}$. Furthermore, if M and $\tilde{M}$ are simply connected and geodesically complete, then there exists a unique global isometry Ψ from M onto $\tilde{M}$ so that $\Psi(P) = \tilde{P}$ and $\Psi_*(P) = \psi$.

Proof. Fix a basis $\{e_1, \dots, e_m\}$ for $T_P M$ and let $\vec{x} = \exp_P(x^1 e_1 + \dots + x^m e_m)$ be geodesic coordinates near P. Fix $v \in \mathbb{R}^m$. The straight line $\sigma(s) := s \cdot v$ in the direction of v through the origin is a geodesic in M. Let $\{E_1(s), \dots, E_m(s)\}$ be the parallel frame field along σ with initial condition $E_i(0) = e_i$. Since $\nabla \mathcal{R} = 0$, the components of the curvature tensor relative to the frame $\{E_i\}$ are constant along σ:

$$\begin{aligned}\partial_s R(E_i, E_j, E_k, E_\ell) &= (\nabla_{\dot\sigma} R)(E_i, E_j, E_k, E_\ell) + R(\nabla_{\dot\sigma} E_i, E_j, E_k, E_\ell)\\ &+ R(E_i, \nabla_{\dot\sigma} E_j, E_k, E_\ell) + R(E_i, E_j, \nabla_{\dot\sigma} E_k, E_\ell) + R(E_i, E_j, E_k, \nabla_{\dot\sigma} E_\ell) = 0\,.\end{aligned}$$

If Y is a vector field along σ, let $\mathcal{J}(\dot\sigma) : Y :\to \mathcal{R}(Y, \dot\sigma)\dot\sigma$ be the *Jacobi operator*; Y is said to be a *Jacobi vector field* if $\ddot{Y} + \mathcal{R}(Y, \dot\sigma)\dot\sigma = 0$. Let $T(s,t) := s(v + tw)$ be a geodesic spray. By Lemma 3.16 in Book I, the associated variational vector $Y(s) = \partial_t T(s,t)|_{t=0} = sw$ is a Jacobi vector field along σ. Since the components of the curvature tensor are constant relative to the frame E_i, Y satisfies a constant coefficient ordinary differential equation relative to the moving frame $\{E_i(s)\}$ with initial condition $Y(0) = 0$ and $\dot{Y}(0) = w$. Consequently, the components of Y are completely determined by $Y(0) = 0$ and $\dot{Y}(0) = w$. Therefore, $g(sw, sw)(\sigma(s))$ is determined by the data $\{s, g(e_i, e_j), (w, e_i), R_{ijk\ell}(0)\}$ where (w, e_i) is the usual Euclidean inner product. Let $\tilde{e}_i = \psi e_i$. We perform the same construction on $\tilde{M}$ where $\tilde\sigma = \Psi\sigma$. As

$$\Psi(\exp_P(s(v + tw))) = \exp_{\tilde P}(s(\psi v + t\psi w))\,,$$

we have $\Psi_*(Y) = \tilde{Y}$. Since $g(Y(s), Y(s))(\sigma(s)) = \tilde{g}(\tilde{Y}(s), \tilde{Y}(s))(\tilde\sigma(s))$, Ψ is an isometry away from 0; at 0, Ψ is an isometry by construction.

Assume additionally that (M, g) and $(\tilde{M}, \tilde{g})$ are geodesically complete and simply connected. Let $\sigma(t) := \exp_P(tv)$ and let $\tilde\sigma(t) := \exp_{\tilde P}(t\psi v)$ be geodesics on M and on $\tilde{M}$, respectively. Fix $0 < T < \infty$. Since $\sigma([0,T])$ and $\tilde\sigma([0,T])$ are compact, we can cover $\sigma([0,T])$ and $\tilde\sigma([0,T])$ by a finite number of open sets where the inverse of the exponential map is a diffeomorphism. We can then recursively apply the local construction described above to extend Ψ as a local isometry along $\sigma(s)$ so $\Psi \circ \sigma = \tilde\sigma$. We can join any two points of M by broken geodesics and extend Ψ along a path of broken geodesics. The assumption that M is simply connected then implies there is no problem with holonomy and defines $\Psi : M \to \tilde{M}$ so that $\Psi^*(\tilde{g}) = g$. We apply a similar construction to ψ^{-1} to construct Ψ^{-1} and see Ψ is a global isometry with the desired properties. By Lemma 7.4, Ψ is uniquely determined by the conditions $\Psi(P) = \tilde{P}$ and $\Psi_*(P) = \psi$. □

We remark that the assumption that M and $\tilde{M}$ are simply connected in Lemma 7.15 is necessary. If $\mathcal{M} := (\mathbb{T}^2, dx^2 + dy^2)$ and if $\tilde{\mathcal{M}} := (\mathbb{R}^2, dx^2 + dy^2)$, then $\mathcal{M}$ and $\tilde{\mathcal{M}}$ are complete flat Riemannian manifolds which are locally isometric; there is, however, no global isometry from $\mathcal{M}$ onto $\tilde{\mathcal{M}}$. Similarly, the assumption that $\mathcal{M}$ and $\tilde{\mathcal{M}}$ are complete in Lemma 7.15 is necessary. If $\mathcal{M} := ((0,2), dx^2)$ and if $\tilde{\mathcal{M}} := ((0,1), dx^2)$, then $\mathcal{M}$ and $\tilde{\mathcal{M}}$ are simply connected flat Riemannian manifolds which are locally isometric. But there is no global isometry from $\mathcal{M}$ onto $\tilde{\mathcal{M}}$.

7.5.1 THE GEODESIC INVOLUTION. Let (M, g) be a pseudo-Riemannian manifold. The *geodesic involution* S_P about a point P of M is defined by setting

$$S_P(Q) := \exp_P\{-\log_P(Q)\}\,. \tag{7.5.a}$$

The map S_P is a local diffeomorphism defined near P which is characterized by the property that $(S_P\sigma)(t) = \sigma(-t)$ for any geodesic σ through P.

Lemma 7.16 Let (M, g) be a connected pseudo-Riemannian manifold.

1. The following assertions are equivalent and if any is satisfied, then (M, g) is said to be a *local symmetric space*:
   (a) The geodesic symmetry S_P is a local isometry defined near P for all $P \in M$.
   (b) $\nabla\mathcal{R} = 0$. This means that the curvature operator is parallel.
   (c) If X, Y, and Z are parallel vector fields along a curve γ, then $\mathcal{R}(X, Y)Z$ is a parallel vector field along γ.
2. If (M, g) is a local symmetric space, then (M, g) is locally homogeneous.

Proof. Let $\vec{x}$ be a system of geodesic coordinates on M centered at P. Then $S_P(\vec{x}) = -\vec{x}$. Let $A_i = \partial_{x_{v_i}}$ be coordinate vector fields; $(S_P)_*(A_i) = -A_i$. If S_P is a local isometry, then we have that $S_P^*(\nabla R) = \nabla R$. Consequently,

$$\begin{aligned}(-1)^5\nabla_{A_1}R(A_2, A_3, A_4, A_5)(0) &= \nabla R_{-A_1}(-A_2, -A_3, -A_4, -A_5)(0)\\ &= \nabla_{A_1}R(A_2, A_3, A_4, A_5; A_1)(0)\,.\end{aligned}$$

This implies that $\nabla R(0) = 0$ and shows that Assertion 1-a implies Assertion 1-b. Conversely, if Assertion 1-b holds, then we may apply Lemma 7.15 with $M = \tilde{M}$, $P = \tilde{P}$, and $\psi = -\operatorname{Id}$ to see that S_P is a local isometry and show that Assertion 1-a holds. Let $\{e_i\}$ be a parallel frame along a curve γ. The equivalence of Assertion 1-b and Assertion 1-c follows from the identity $(\nabla_{\dot{\gamma}}R)(e_i, e_j, e_k, e_\ell) = \dot{\gamma}R(e_i, e_j, e_k, e_\ell)$.

Assume that S_P is a local isometry for all $P \in M$. Let γ be a curve from a point Q_1 to a point Q_2 of M. Since range$\{\gamma\}$ is compact, we may cover range$\{\gamma\}$ by a finite number of open sets so that any two points in one of the open sets can be joined by unique shortest geodesic and so that the geodesic involution about the midpoint of such a geodesic is a local isometry interchanging the two end points. Composing such geodesic involutions then yields a local isometry interchanging Q_1 and Q_2. □

7.6 THE GLOBAL GEOMETRY OF SYMMETRIC SPACES

We say that a connected pseudo-Riemannian manifold (M, g) is a *symmetric space* if the geodesic involution S_P of Equation (7.5.a) extends for any $P \in M$ to a global isometry of M. The following is an immediate consequence of Lemma 7.15.

Lemma 7.17 If (M, g) is a simply connected geodesically complete local symmetric space, then (M, g) is a symmetric space.

The assumption that (M, g) is simply connected is essential in Lemma 7.17. Let the group of nth roots of unity $\mathbb{Z}_n$ act by complex multiplication without fixed points isometrically on $S^{2k-1} \subset \mathbb{C}^k$. Give the quotient manifold $L(n, k) := S^{2k-1}/\mathbb{Z}_n$ the induced metric so that the covering projection from S^{2k-1} to $L(n; k)$ is a local isometry; $L(n, k)$ is an example of a *lens space*. Then $L(n; k)$ is locally symmetric and geodesically complete. If $P = (1, 0, \dots, 0)$, then $S_P(\vec{x}) = (x^1, -x^2, \dots, -x^{2k})$. Since S_P does not commute with the action of $\mathbb{Z}_n$ for $n \geq 3$, $L(n; k)$ is not globally symmetric.

Lemma 7.18 If (M, g) is a symmetric space, then the map $\Phi : (P, Q) \to S_P(Q)$ is a smooth map from $M \times M$ to M.

Proof. Let $B_\epsilon(Q)$ be the open ball of radius ϵ about Q relative to some distance function on M defining the topology. Let $P, \tilde{P} \in M$. We must show that the map Φ is smooth near $(P, \tilde{P})$. Let $\sigma : [0, T] \to M$ be a curve from P to $\tilde{P}$ in M. Choose $\epsilon > 0$ so $B_{2\epsilon}(R) \subset \text{domain}\{\log_R\}$ for any $R \in \text{range}\{\sigma\}$, i.e., so $\exp_R$ is a diffeomorphism from $\log_P(B_{2\epsilon}(R))$ to $B_{2\epsilon}(R)$. Choose $\delta > 0$ so that if $|s - t| < \delta$, then $d(\sigma(s), \sigma(t)) < \epsilon$ and so that $T = n\delta$ for some $n \in \mathbb{N}$. Let $R_j := \sigma(j\delta)$ for $1 \leq j \leq n$. If $Q \in B_\epsilon(P)$ and if $R \in B_\epsilon(P)$, then $R \in B_{2\epsilon}(Q)$ and

$$\Phi(Q, R) = S_Q(R) = \exp_Q(-\log_Q(R)) .$$

Consequently, Φ is smooth on $B_\epsilon(P) \times B_\epsilon(P)$. We have $P = R_0$. Choose $0 \leq j \leq n$ maximal so Φ is smooth on $B_\epsilon(P) \times B_\epsilon(R_j)$. Suppose $j < n$; we argue for a contradiction. Let Q belong to $B_\epsilon(P)$. As S_Q is a global isometry of M, S_Q commutes with the exponential map. Let R belong to $B_{2\epsilon}(R_j)$. Then

$$\Phi(Q, R) = S_Q(R) = S_Q(\exp_{R_j}(\log_{R_j}(R))) = \exp_{S_Q(R_j)}(S_Q(R_j)_*(\log_{R_j}(R)))$$

so Φ is smooth on $B_\epsilon(P) \times B_{2\epsilon}(R_j)$. Since $d(R_{j+1}, R_j) < \epsilon$, $B_\epsilon(R_{j+1}) \subset B_{2\epsilon}(R_j)$ and Φ is smooth on $B_\epsilon(P) \times B_\epsilon(R_{j+1})$. This contradicts the maximality of j. Consequently, $j = n$ and, since $R = R_n$, Φ is smooth on $B_\epsilon(P) \times B_\epsilon(R)$. □

Let (M, g) be a symmetric space. If σ is a non-constant geodesic, then the associated *transvection* is defined by setting:

$$\Psi_t = \Psi_{t,\sigma} := S_{\sigma(\frac{1}{2}t)} S_{\sigma(0)} \in \mathcal{I}(M, g) . \tag{7.6.a}$$

The following result provides a partial converse to Lemma 7.17 by showing that any symmetric space is geodesically complete.

Lemma 7.19 Let (M, g) be a symmetric space. Let $\Psi_t := S_{\sigma(\frac{1}{2}t)}S_{\sigma(0)}$ be the transvection of a geodesic σ. Then $\Psi_t(\sigma(s)) = \sigma(t+s)$, (M, g) is geodesically complete, and Ψ_t is defined for all $t \in \mathbb{R}$. Either σ is injective or σ is simply periodic. If X is a parallel vector field along σ, then $(\Psi_t)_*X = X$. The map $(\Psi_t)_*(\sigma(s)) : T_{\sigma(s)}M \to T_{\sigma(t+s)}(M)$ is given by parallel translation along σ. Finally, $\{\Psi_t\}_{t\in\mathbb{R}}$ is a smooth one-parameter group of isometries in the group of isometries $\mathcal{I}(M, g)$.

Proof. Since $S_{\sigma(t)}\sigma(s) = \sigma(2t - s)$ for t and s small, the transvection Ψ_t of Equation (7.6.a) satisfies:

$$\Psi_t\sigma(s) = S_{\sigma(\frac{1}{2}t)}S_{\sigma(0)}\sigma(s) = S_{\sigma(\frac{1}{2}t)}\sigma(-s) = \sigma(t+s)\,.$$

We may use the transvections Ψ_t to translate the geodesic σ and to extend σ (which a-priori is only defined on $(-\epsilon, \epsilon)$ for some $\epsilon > 0$) to $(-\infty, \infty)$. Consequently, (M, g) is geodesically complete. If $\sigma(0) = \sigma(b)$ for some b, then $\sigma(t) = \Psi_t\sigma(0) = \Psi_t\sigma(b) = \sigma(b+t)$ for any t so σ is simply periodic. Let X be a parallel vector field along σ. Since $S_{\sigma(t)}$ is an isometry, it preserves parallel vector fields. Since $(S_{\sigma(t)})_*(\sigma(t)) = -1$ on $T_{\sigma(t)}M$, $S_{\sigma(t)}X = -X$ for any t. Therefore, $(\Psi_t)_*X = X$ as we have changed the sign twice. Furthermore,

$$(\Psi_t)_*(\sigma(s))X(\sigma(s)) = X(\sigma(s+t))$$

so the action of $(\Psi_t)_*$ mapping $T_{\sigma(s)}$ to $T_{\sigma(t)}$ is given by parallel translation. By Lemma 7.4, an isometry is determined by its value and its derivative at a single point. We have

$$\Psi_t\Psi_s(0) = \Psi_t\{\sigma(s)\} = \sigma(s+t)\,.$$

Furthermore, first parallel translating from $\sigma(0)$ to $\sigma(s)$ and then parallel translating from $\sigma(s)$ to $\sigma(t)$ is parallel translating from $\sigma(0)$ to $\sigma(s+t)$. Consequently, $\Psi_t\Psi_s = \Psi_{t+s}$ and Ψ_t is a 1-parameter family. By Lemma 7.18, the map $(t, Q) \to S_{\sigma(\frac{1}{2}t)}S_{\sigma(0)}$ is smooth in (t, Q). Consequently, by Theorem 7.6, Ψ_t is a smooth 1-parameter family of the group of isometries $\mathcal{I}(M, g)$.
□

Theorem 7.20 *Let (M, g) be a simply connected symmetric space. Then any local Killing vector field on M extends to a global Killing vector field on M, the vector space of Killing vectors is the Lie algebra of $\mathcal{I}(M, g)$, and (M, g) is a homogeneous space.*

Proof. By Lemma 7.19, (M, g) is geodesically complete. We adopt the notation of Theorem 7.11. Let $W = TM \oplus (T^*M \otimes TM)$. If X is a local Killing vector field, let

$$\mathfrak{S}_X(P) := X(P) \oplus \nabla X(P) \in W_P$$

and let $\mathfrak{K}_P = \cup_X \mathfrak{S}_P(X)$ be a linear subspace of W_P. By Lemma 7.16, (M, g) is locally homogeneous. Thus, $\dim\{\mathfrak{K}_P\}$ is constant on M. By Theorem 7.11, since (M, g) is simply connected and geodesically complete, every local Killing vector field on M extends to a global Killing vector field.

By Lemma 7.9, since (M, g) is simply connected and geodesically complete, every global Killing vector field defines a 1-parameter family of isometries and the Lie algebra of global Killing vector fields is the Lie algebra of the isometry group $\mathcal{I}(M, g)$. Fix $P \in M$. The transvections defined by the geodesics through P define m linearly independent Killing vector fields at P. Consequently, by Lemma 7.10, (M, g) is homogeneous. □

Theorem 7.21 *Let (M, g) be a pseudo-Riemannian manifold. Let N be a connected smooth connected submanifold of M so that $g|_N$ is non-degenerate and so that $(N, g|_N)$ is totally geodesic. If (M, g) is a local symmetric space, then $(N, g|_N)$ is a local symmetric space. If (M, g) is a symmetric space, then $(N, g|_N)$ is a symmetric space.*

Proof. Assume that (M, g) is a local symmetric space and that N is a totally geodesic non-degenerate submanifold of M. Let $P \in N$. Since N is totally geodesic, the geodesic symmetry S_P of M preserves N and hence N is a local symmetric space. If M is a symmetric space, then S_P is globally defined. Let

$$\tilde{N} := \{Q \in N : S_P(Q) \in N \text{ and } (S_P)_*(Q) : T_Q N \to T_{S_P(Q)}(N)\}.$$

Then $\tilde{N}$ is a closed set. As N is totally geodesic, it follows that $\tilde{N}$ is open. Since N is connected, $\tilde{N} = N$ and the restriction of S_P to N provides the requisite geodesic involution on N. □

Let (M, g) be a symmetric space. By Theorem 7.20, the connected component of the isometry group $\mathcal{I}_0(M, g)$ acts transitively on M. Let $H = H_P \subset \mathcal{I}_0(M, g)$ be the isotropy subgroup of the connected component of the identity of the isometry group of (M, g). By Theorem 7.2, the natural map $\theta \to \theta \cdot P$ is a diffeomorphism from $\mathcal{I}_0(M, g)/H$ to M.

Note that the geodesic involution S_P belongs to $\mathcal{I}(M, g)$ but that S_P need not belong to $\mathcal{I}_0(M, g)$. If we take $(M, g) = (\mathbb{T}^2, dx^2 + dy^2)$, then S_P^* acts non-trivially on the de Rham cohomology of the torus and hence S_p is not in the identity component of the isometry group. On the other hand, if we take $(M, g) = (\mathbb{R}^2, dx^2 + dy^2)$ and let $T_\theta(P)$ be rotation through an angle θ about some point $P \in \mathbb{R}^2$, $\Theta_t(P)$ is a smooth 1-parameter family of isometries about P with $\Theta_\pi = S_P$ and $S_P \in \mathcal{I}_0(M, g)$.

Define an involution $\sigma = \sigma_P$ of $\mathcal{I}(M, g)$ by setting

$$\sigma(\theta) := S_P \circ \theta \circ S_P . \tag{7.6.b}$$

Since $\mathcal{I}_0(M, g)$ is a normal subgroup of $\mathcal{I}(M, g)$, it is preserved by σ. Let

$$G^\sigma(M, g) := \{\theta \in \mathcal{I}_0(M, g) : \sigma(\theta) = \theta\}.$$

This Lie group need not be connected. Let $G_0^\sigma(M, g)$ be the connected component of the identity of the Lie group G; in Example 7.6.1 we exhibit (G, σ) so that $G^\sigma \neq G_0^\sigma$.

Lemma 7.22 If (M, g) is a symmetric space, then $G_0^\sigma(M, g) \subset H(M, g) \subset G^\sigma(M, g)$.

Proof. If $\theta \in H$, let $\psi := \theta^{-1} S_P \theta S_P$. Since $S_P(P) = P$ and $\theta(P) = P$, $\psi(P) = P$. We use the chain rule to see that $\psi_*(P) = (\theta_*(P)^{-1})(-\operatorname{Id})(\theta_*(P))(-\operatorname{Id}) = \operatorname{Id}$. Consequently, by Lemma 7.4, $\psi = \operatorname{Id}$ and $s_P \theta s_P = \theta$. This shows

$$H \subset G^\sigma .$$

We show next that $G_0^\sigma \subset H$. Since G_0^σ is connected, G_0^σ is generated by the 1-parameter subgroups of G_0^σ and it suffices to show that $\exp(t\xi) \in H$ for any ξ in the Lie algebra of G_0^σ. We have

$$S_P \exp(t\xi) S_P^{-1} = \exp(t S_P \xi S_P) = \exp(t \operatorname{Ad}(S_P)\xi) .$$

This shows that the Lie algebra of G_0^σ consists of the fixed points of the involution under the adjoint action. Consequently, if $\xi \in \mathfrak{g}(G^\sigma)$, then $S_P \exp(t\xi) = \exp(t\xi) S_P$. Therefore, $S_P \exp(t\xi) P = P$. Since P is an isolated fixed point of S_P, this implies $\exp(t\xi) P = P$ and hence $\exp(t\xi) \in H$ as desired. □

We now pass to the algebraic level.

Theorem 7.23 *Let σ be a group homomorphism of a connected Lie group G with $\sigma^2 = \operatorname{Id}$. Let $G^\sigma := \{\theta \in G : \sigma(\theta) = \theta\}$; G^σ is a closed Lie subgroup of G. Let G_0^σ be the connected component of the identity in G^σ. Let H be a closed subgroup of G with $G_0^\sigma \subset H \subset G^\sigma$. Then every G-invariant pseudo-Riemannian metric on $M = G/H$ gives M the structure of a symmetric space. Let π be the natural projection from G to G/H. Then $\pi(\sigma(g)) = S_P \pi(g)$ where S_P is the geodesic involution based at $P = \pi(\operatorname{Id})$.*

Proof. Let $M = G/H$ be given a G-invariant pseudo-Riemannian metric. Let $P = \pi(\operatorname{Id})$ be the base point. Since the involution σ fixes H, we have

$$\sigma(g_1 H) = \sigma(g_1)\sigma(H) = \sigma(g_1) H .$$

If $\pi(g_1) = \pi(g_2)$, then $g_1 H = g_2 H$ and, thus, $\pi(\sigma(g_1)) = \pi(\sigma(g_2))$. Consequently, $S_P(\pi g) := \pi(\sigma(g))$ is well-defined. The existence of local sections to the submersion π implies that S_P is smooth. Clearly, $S_P(P) = P$. Since $\sigma^2 = \operatorname{Id}$, $S_P^2 = \operatorname{Id}$. Thus, S_P is a smooth involution of M.

Let $\mathfrak{g}$ and $\mathfrak{h}$ be the Lie algebras of G and H, respectively. We then have that $\mathfrak{g}$ is the direct sum of $\mathfrak{h}$ and the subspace $\mathfrak{m} = \{X \in \mathfrak{g} : d\sigma(X) = -X\}$. We argue as follows to see this. If $X \in \mathfrak{g}$, decompose $X = X_\mathfrak{h} + X_\mathfrak{m}$, where $X_\mathfrak{h} = \frac{1}{2}(X + d\sigma(X))$ and $X_\mathfrak{m} = \frac{1}{2}(X - d\sigma(X))$. Since σ is involutive, so is $d\sigma$, and, therefore, $d\sigma(X_\mathfrak{h}) = X_\mathfrak{h}$ and $d\sigma(X_\mathfrak{m}) = -X_\mathfrak{m}$. It follows that $\mathfrak{g} = \mathfrak{h} + \mathfrak{m}$, which is a direct sum since $\mathfrak{h} \cap \mathfrak{m} = 0$. Next, note that $S_P(P) = P$ and if $y \in T_P M$ the previous argument implies that there exists $Y \in \mathfrak{g}$ such that $d\sigma(Y) = -Y$ and $d\pi(Y) = y$. We show that $dS_P = -\operatorname{Id}$ by computing:

$$dS_P(y) = dS_P(d\pi(Y)) = d\pi(d\sigma(Y)) = d\pi(-Y) = -y\,.$$

Finally, let $\langle\,,\,\rangle$ be a G-invariant metric tensor on $M = G/H$ and associated to each element g in G consider the translation $\tau_g : G/H \to G/H$ given by $\tau_g(g_1H) = gg_1H$. For $a \in G$, we have

$$\begin{aligned} S_P(\tau_g(\pi(a))) &= S_P(\pi(ga)) = \pi(\sigma(ga)) = \pi(\sigma(g)\sigma(a)) \\ &= \tau_{\sigma(g)}(\pi(\sigma(a))) = \tau_{\sigma(g)}(S_P(\pi(a)))\,, \end{aligned}$$

so $\tau_{\sigma(g)} = S_P\tau_g S_P$. As a consequence, if $v \in T_gM$ and $v_P = d\tau_{g^{-1}}(v) \in T_PM$, we get

$$\begin{aligned} \langle dS_P(v), dS_P(v)\rangle &= \langle dS_P d\tau_g(v_P), dS_P d\tau_g(v_P)\rangle \\ &= \langle d\tau_{\sigma(g)}dS_P(v_P), d\tau_{\sigma(g)}dS_P(v_P)\rangle \\ &= \langle dS_P(v_P), dS_P(v_P) = \langle -v_P, -v_P\rangle = \langle v, v\rangle \end{aligned}$$

This shows that the map S_P is an isometry with respect to $\langle\,,\,\rangle$. If a homogeneous space has a global symmetry at P, then $\tau S_P \tau^{-1}$ is a global symmetry at any point $\tau(P)$. This completes the proof. □

By Lemma 7.22, every symmetric space arises from the construction of Theorem 7.23. Consequently, Theorem 7.23 reduces the study of symmetric spaces to a group theoretic problem. We refer to Helgason [30] for a further discussion in this regard; this is a vast field.

Let G be a compact Lie group and let $\sigma \in G$ be involutive. Suppose that H is a closed subgroup of G with $(G_0)_0^\sigma \subset H \subset (G_0)^\sigma$. Set $M := G_0/H$. By Lemma 7.12, there exists a Riemannian metric ds^2 on M so G acts by isometries; (M, ds^2) is then a symmetric space.

7.6.1 EXAMPLE. Let $G = \mathrm{SO}(m)$ and let $\sigma : \mathrm{SO}(m) \to \mathrm{SO}(m)$ be given by:

$$\sigma(A) := \begin{pmatrix} \mathrm{Id}_k & 0 \\ 0 & -\mathrm{Id}_{m-k} \end{pmatrix} A \begin{pmatrix} \mathrm{Id}_k & 0 \\ 0 & -\mathrm{Id}_{m-k} \end{pmatrix}^{-1}.$$

Then $G^\sigma = \{\mathrm{O}(k) \times \mathrm{O}(m-k)\} \cap \mathrm{SO}(m)$ and $G_0^\sigma = \mathrm{SO}(k) \times \mathrm{SO}(m-k)\}$; there is a natural double cover $\mathbb{Z}_2 \to G^\sigma \to G_0^\sigma$. Taking $H = G^\sigma$ yields the Grassmann manifold $Gr_k(\mathbb{R}^m)$ of k planes in $\mathbb{R}^m$; taking $H = G_0^\sigma$ yields the Grassmann manifold $\mathrm{Gr}_k^+(\mathbb{R}^m)$ of oriented k planes in $\mathbb{R}^m$. The double cover $\mathbb{Z}_2 \to \mathrm{Gr}_k^+(\mathbb{R}^m) \to \mathrm{Gr}_k(\mathbb{R}^m)$ arises by forgetting the orientation. Taking $k = 1$ yields the sphere S^{m-1} and real projective space $\mathbb{RP}^{m-1}$, respectively.

7.6.2 EXAMPLE. A Lie group G is a symmetric space determined by the symmetric pair $(G \times G, \Delta_G)$, where Δ_G is the diagonal subgroup $\Delta_G = \{(g, g) \in G \times G : g \in G\}$ and where the involution σ interchanges the two factors, i.e., $\sigma(g_1, g_2) = (g_2, g_1)$; we may take any left-invariant pseudo-Riemannian metric on G.

CHAPTER 8

Other Cohomology Theories

In Chapter 8, we show that de Rham cohomology is isomorphic to singular cohomology, PL cohomology, and sheaf cohomology. We begin in Section 8.1 by presenting some standard results in homological algebra that we used to complete the proof of Theorem 5.2 by establishing the Mayer–Vietoris sequence in de Rham cohomology and by showing that de Rham cohomology is a homotopy functor. In Section 8.2, we discuss simplicial theory. In Section 8.3, we discuss singular (i.e., topological or simply TP) cohomology $H_*^{\mathrm{TP}}(\cdot)$ and $H^*_{\mathrm{TP}}(\cdot)$. In Section 8.4, we treat sheaf cohomology.

8.1 HOMOLOGICAL ALGEBRA

We begin with some basic definitions.

8.1.1 THE LANGUAGE OF CATEGORY THEORY. The reader may safely bypass this material as the terminology involved will be, for the most part, clear by context subsequently; it is presented for the sake of completeness. Let $\mathfrak{C}$ be a category. We shall work in the category of simplicial complexes (the PL category), the category of topological spaces (the TP category), and the category of smooth manifolds (the C^∞ category). Let $\mathcal{H}$ be a contravariant functor from the category $\mathfrak{C}$ to the category of real or complex vector spaces. If X is an element of $\mathfrak{C}$, let $\mathcal{H}(X)$ be the associated vector space. If $f : X \to Y$ is a *morphism* in the category, then we have an associated map $f^* : \mathcal{H}(Y) \to \mathcal{H}(X)$ so that

$$\mathrm{Id}^* = \mathrm{Id} \quad \text{and} \quad (f \circ g)^* = g^* \circ f^* .$$

The morphisms in the category of smooth manifolds are the smooth maps, the morphisms in the category of topological spaces are the continuous maps, and the morphisms in the simplicial category are the simplicial maps. As discussed in Chapter 5, de Rham cohomology is a functor which is defined on the smooth category. Simplicial cohomology (see Section 8.2) is defined on the PL-category, singular cohomology (see Section 8.3) is defined on the topological category, and sheaf cohomology (see Section 8.4) is defined on the topological category. We have natural inclusions of the simplicial and smooth categories in the topological category. If $\mathcal{H}_1$ is a functor on a category $\mathfrak{C}_1$, if $\mathcal{H}_2$ is a functor on a category $\mathfrak{C}_2$, and if $i : \mathfrak{C}_1 \to \mathfrak{C}_2$ is a map between categories, then a *natural transformation of functors* τ is a linear map τ_X from $\mathcal{H}_1(X)$ to $\mathcal{H}_2(i(X))$ for each element X of $\mathfrak{C}_1$ so that if $f : X \to Y$ is a morphism in $\mathfrak{C}_1$, then we have a commutative diagram:

$$\begin{array}{ccc} \mathcal{H}_1(Y) & \xrightarrow{\tau_Y} & \mathcal{H}_2(i(Y)) \\ \downarrow f_1^* & \circ & \downarrow f_2^* \\ \mathcal{H}_1(X) & \xrightarrow{\tau_X} & \mathcal{H}_2(i(X)) \end{array} .$$

8.1.2 CHAIN COMPLEXES AND COCHAIN COMPLEXES. A *cochain complex* $\mathcal{A}$ is a collection of real vector spaces A_i together with linear maps $d_i^A : A_i \to A_{i+1}$ where the A_i are real (or possibly complex) vector spaces and where the d_i^A are linear maps so that $d_i^A d_{i-1}^A = 0$; it may be represented by the diagram:

$$0 \to A_0 \xrightarrow{d_0^A} A_1 \to \cdots \to A_{p-1} \xrightarrow{d_{p-1}^A} A_p \xrightarrow{d_p^A} A_{p+1} \cdots .$$

The *cohomology* of the cochain complex $\mathcal{A}$ is defined to be:

$$H^i(\mathcal{A}) := \frac{\ker\{d_i^A : A_i \to A_{i+1}\}}{\text{range}\{d_{i-1}^A : A_{i-1} \to A_i\}} .$$

If $a \in \ker\{d_i^A\}$, we let $[a] \in H^i(\mathcal{A})$ denote the corresponding element in cohomology. A *cochain map* $\mathcal{T}$ from a cochain complex $\mathcal{A} = (A_i, d_i^A)$ to a cochain complex $\mathcal{B} = (B_i, d_i^B)$ is a collection of linear maps $\mathcal{T} = (T_i : A_i \to B_i)$ so $d_i^B T_i = T_{i+1} d_i^A$. We represent this as:

$$\begin{array}{ccccccccc} 0 & \to & A_0 & \xrightarrow{d_0^A} & A_1 & \to \cdots \to & A_i & \xrightarrow{d_i^A} & A_{i+1} \cdots \\ & & \downarrow T_0 & \circ & \downarrow T_1 & \cdots & \downarrow T_i & \circ & \downarrow T_{i+1} \\ 0 & \to & B_0 & \xrightarrow{d_0^B} & B_1 & \to \cdots \to & B_i & \xrightarrow{d_i^B} & B_{i+1} \cdots \end{array} .$$

Let $\mathcal{S}$ and $\mathcal{T}$ be cochain maps from a cochain complex $\mathcal{A}$ to a cochain complex $\mathcal{B}$. We say that $\mathcal{R} = \{R_i : A_i \to B_{i-1}\}$ is a *cochain homotopy* between $\mathcal{S}$ and $\mathcal{T}$ if

$$d_{i-1}^B R_i + R_{i+1} d_i^A = T_i - S_i .$$

There is always a question about notation; we will write the indices on the vector spaces down, the maps will go to the right or down, and the indices on the cohomology groups will be up. When no confusion is likely to result, we suppress superscripts and subscripts and set $d = d_i^A$ and $T = T_i$.

Lemma 8.1 Let $\mathcal{T}$ be a cochain map from a cochain complex $\mathcal{A}$ to a cochain complex $\mathcal{B}$. Then the map $T_* : [a] \to [Ta]$ is a well-defined map in cohomology from $H^p(\mathcal{A})$ to $H^p(\mathcal{B})$. If $\mathcal{T}$ and $\mathcal{S}$ are cochain homotopic, then $T_* = S_*$.

Proof. Let $\theta \in H^p(\mathcal{A})$. Find $a_p \in A_p$ so $[a_p] = \theta$. Since $da_p = 0$, $dTa_p = Tda_p = 0$ so $[Ta_p]$ is well-defined in $H^p(\mathcal{B})$. If $[\tilde{a}_p] = \theta$ is another possible representative of θ, then there exists $a_{p-1} \in A_{p-1}$ so $da_{p-1} = a_p - \tilde{a}_p$. Consequently, $T(a_p - \tilde{a}_p) = Tda_{p-1} = dTa_{p-1}$. This shows that $[Ta_p] = [T\tilde{a}_p]$ in $H^p(\mathcal{B})$ and T_* is well-defined in cohomology. Let $\mathcal{T}$ and $\mathcal{S}$ be chain homotopic. As $da_p = 0$,

$$[Sa_p - Ta_p] = [dRa_p + Rda_p] = [dRa_p + 0] = 0$$

and, consequently, $S_* = T_*$ as maps from $H^p(\mathcal{A})$ to $H^p(\mathcal{B})$. □

8.1.3 SHORT EXACT SEQUENCES. We say that $0 \to A \xrightarrow{\alpha} B \xrightarrow{\beta} C \to 0$ is a *short exact sequence* of vector spaces if the maps α and β are linear, if α is injective, if β is surjective, and if $\ker\{\beta\} = \text{range}\{\alpha\}$. We say that a cochain complex $\mathcal{A}$ is a *long exact sequence* if $H^i(\mathcal{A}) = 0$ for all i, i.e., if $\ker\{d_i^{\mathcal{A}}\} = \text{range}\{d_{i-1}^{\mathcal{A}}\}$. Massey [40] (p. 185) states: "Apparently exact sequences were introduced into algebraic topology by Witold Hurewicz in 1941" (see [34]). We now discuss the *combinatorial Laplacian.* The following is a useful observation that is the extension of Theorem 5.13 to the setting at hand.

Lemma 8.2 Let $\mathcal{A}$ be a cochain complex where each A_i is a finite-dimensional vector space. Put a Hermitian inner product $\langle \cdot, \cdot \rangle$ on each A_i and let $\delta_i : A_i \to A_{i-1}$ be the adjoint map. Let $\Delta := \delta d + d\delta$ be the associated Laplacian. Then:

1. $\ker\{\Delta\} = \ker\{d\} \cap \ker\{\delta\}$, $A_i = \ker\{\Delta_i\} \oplus \text{range}\{d_{i-1}\} \oplus \text{range}\{\delta_i\}$ is a direct orthogonal decomposition, and d is an isomorphism from range$\{\delta\}$ to range$\{d\}$. Finally, the map which sends a to $[a]$ defines an isomorphism from $\ker\{\Delta_p\}$ to $H^p(\mathcal{A})$.

2. $\sum_i (-1)^i \dim\{A_i\} = \sum_i \dim\{H^i(\mathcal{A})\}$. In particular, if $\mathcal{A}$ is a long exact sequence, then $\sum_i (-1)^i \dim\{H^i(\mathcal{A})\} = 0$.

Proof. We suppress indices for the moment. If $\Delta a = 0$, then

$$0 = \langle \Delta a, a \rangle = \langle \delta d a, a \rangle + \langle d\delta a, a \rangle = \langle da, da \rangle + \langle \delta a, \delta a \rangle = \|da\|^2 + \|\delta a\|^2 .$$

This shows $\ker\{\Delta\} = \ker\{d\} \cap \ker\{\delta\}$. As Δ is self-adjoint, we have an orthogonal direct sum decomposition

$$A = \ker\{\Delta\} \oplus \text{range}\{\Delta\} = \ker\{\Delta\} \oplus \{\text{range}\{d\delta\} + \text{range}\{\delta d\}\} . \tag{8.1.a}$$

Since $\langle da, \delta\tilde{a} \rangle = \langle dda, \tilde{a} \rangle = 0$, range$\{d\}$ is perpendicular to range$\{\delta\}$. Thus, the decomposition of Equation (8.1.a) is an orthogonal direct sum decomposition. If $d\delta a_2 = 0$, then

$$0 = \langle d\delta a_2, a_2 \rangle = \langle \delta a_2, \delta a_2 \rangle$$

so $\delta a_2 = 0$. This implies d is an injective map from range$\{\delta\}$ to range$\{d\}$. Similarly, we have δ is an injective map from range$\{d\}$ to range$\{\delta\}$. Thus, $\dim\{\text{range}\{\delta\}\} = \dim\{\text{range}\{d\}\}$ and d is an isomorphism from range$\{\delta\}$ to range$\{d\}$.

Let $a \in \ker\{\Delta\}$. Then $a \in \ker\{d\}$ so $[a]$ is an element of cohomology. If $[a] = 0$, then a is in the range of d. But $\ker\{\Delta\} \perp \text{range}\{d\}$. Therefore, $a = 0$ so the map $a \to [a]$ is an injective map from $\ker\{\Delta\}$ to cohomology. Conversely, suppose $da = 0$. Decompose $a = a_0 + da_1 + \delta a_2$ where $a_0 \in \ker\{\Delta\}$. Then $0 = d\delta a_2$. This implies that $\delta a_2 = 0$ and that $a = a_0 + da_1$. Consequently, $[a] = [a_0]$. This completes the proof of Assertion 1.

Let $R_i^d := \text{range}\{d_{i-1}\}$ and $R_{i-1}^\delta := \text{range}\{\delta_i\}$. Because d_{i-1} is an isomorphism from range$\{\delta_i\} \subset A_{i-1}$ to range$\{d_{i-1}\} \subset A_i$, we have $\dim\{R_i^d\} = \dim\{R_{i-1}^\delta\}$. Consequently,

$$\begin{aligned}
&\textstyle\sum_i (-1)^i \dim\{H^i(\mathcal{C})\}\\
&\quad = \textstyle\sum_i (-1)^i \dim\{\ker\{\Delta_i\}\} + \sum_i (-1)^i \{\dim\{R_i^d\} - \dim\{R_{i-1}^\delta\}\}\\
&\quad = \textstyle\sum_i (-1)^i \dim\{\ker\{\Delta_i\}\} + \sum_i (-1)^i \dim\{R_i^d\} + \sum_i (-1)^i \dim\{R_i^\delta\}\\
&\quad = \textstyle\sum_i (-1)^i \dim\{A_i\}\,.
\end{aligned}$$

If $\mathcal{A}$ is a long exact sequence, then the cohomology is trivial so the sum vanishes. □

8.1.4 SHORT EXACT SEQUENCE OF CO-CHAIN COMPLEXES. A *short exact sequence of cochain complexes* $0 \to \mathcal{A} \xrightarrow{\alpha} \mathcal{B} \xrightarrow{\beta} \mathcal{C} \to 0$ is a pair of cochain maps α from $\mathcal{A}$ to $\mathcal{B}$ and β from $\mathcal{B}$ to $\mathcal{C}$ so that the column maps form a short exact sequence $0 \to A_p \xrightarrow{\alpha} B_p \xrightarrow{\beta} C_p \to 0$. Consequently, we have a commutative diagram:

$$\begin{array}{cccccccccc}
 & 0 & & 0 & & 0 & & 0 & \\
 & \downarrow & & \downarrow & & \downarrow & & \downarrow & \\
0 \to & A_0 & \xrightarrow{d_0^{\mathcal{A}}} & A_1 & \cdots & A_i & \xrightarrow{d_i^{\mathcal{A}}} & A_{i+1} & \cdots \\
 & \downarrow \alpha_0 & \circ & \downarrow \alpha_1 & \cdots & \downarrow \alpha_i & \circ & \downarrow \alpha_{i+1} & \cdots \\
0 \to & B_0 & \xrightarrow{d_0^{\mathcal{B}}} & B_1 & \cdots & B_i & \xrightarrow{d_i^{\mathcal{B}}} & B_{i+1} & \cdots \\
 & \downarrow \beta_0 & \circ & \downarrow \beta_1 & \cdots & \downarrow \beta_i & \circ & \downarrow \beta_{i+1} & \cdots \\
0 \to & C_0 & \xrightarrow{d_0^{\mathcal{C}}} & C_1 & \cdots & C_i & \xrightarrow{d_i^{\mathcal{C}}} & C_{i+1} & \cdots \\
 & \downarrow & & \downarrow & & \downarrow & & \downarrow & \\
 & 0 & & 0 & & 0 & & 0 &
\end{array}\,.$$

The following result is due to the French mathematician Élie Cartan and the Polish mathematician Samuel Eilenberg [12] (p. 40); it is often called the *Snake Lemma*.

E. Cartan (1869–1951)

S. Eilenberg (1913–1998)

Lemma 8.3 Let $0 \to \mathcal{A} \xrightarrow{\alpha} \mathcal{B} \xrightarrow{\beta} \mathcal{C} \to 0$ be a short exact sequence of cochain complexes. There exists a natural map $\mathfrak{v} : H^{p-1}(\mathcal{C}) \to H^p(\mathcal{A})$, which is called the *connecting homomorphism*, giving rise to a long exact sequence in cohomology:

$$0 \to H^0(\mathcal{A}) \xrightarrow{\alpha_*} H^0(\mathcal{B}) \xrightarrow{\beta_*} H^0(\mathcal{C}) \xrightarrow{\mathfrak{v}} H^1(\mathcal{A}) \xrightarrow{\alpha_*} H^1(\mathcal{B}) \xrightarrow{\beta_*} H^1(\mathcal{C}) \xrightarrow{\mathfrak{v}} H^2(\mathcal{A}) \to \cdots\,.$$

The connecting homomorphism $\mathfrak{v}$ is natural in the sense that a commutative diagram of short exact sequences of cochain complexes

$$\begin{array}{ccccccccc}
0 & \rightarrow & \mathcal{A} & \xrightarrow{\alpha} & \mathcal{B} & \xrightarrow{\beta} & \mathcal{C} & \rightarrow & 0 \\
 & & \downarrow \sigma_{\mathcal{A}} & \circ & \downarrow \sigma_{\mathcal{B}} & \circ & \downarrow \sigma_{\mathcal{C}} & & \\
0 & \rightarrow & \tilde{\mathcal{A}} & \xrightarrow{\tilde{\alpha}} & \tilde{\mathcal{B}} & \xrightarrow{\tilde{\beta}} & \tilde{\mathcal{C}} & \rightarrow & 0
\end{array}$$

gives rise to a commutative diagram of long exact sequences:

$$\begin{array}{l}
0 \rightarrow H^0(\mathcal{A}) \xrightarrow{\alpha_*} H^0(\mathcal{B}) \xrightarrow{\beta_*} H^0(\mathcal{C}) \xrightarrow{\mathfrak{v}} H^1(\mathcal{A}) \xrightarrow{\alpha_*} H^1(\mathcal{B}) \xrightarrow{\beta_*} H^1(\mathcal{C}) \xrightarrow{\mathfrak{v}} H^2(\mathcal{A}) \cdots \\
\quad \downarrow (\sigma_A)_* \circ \downarrow (\sigma_B)_* \circ \downarrow (\sigma_C)_* \circ \downarrow (\sigma_A)_* \circ \downarrow (\sigma_B)_* \circ \downarrow (\sigma_C)_* \circ \downarrow (\sigma_A)_* \cdots \\
0 \rightarrow H^0(\tilde{\mathcal{A}}) \xrightarrow{\tilde{\alpha}_*} H^0(\tilde{\mathcal{B}}) \xrightarrow{\tilde{\beta}_*} H^0(\tilde{\mathcal{C}}) \xrightarrow{\tilde{\mathfrak{v}}} H^1(\tilde{\mathcal{A}}) \xrightarrow{\tilde{\alpha}_*} H^1(\tilde{\mathcal{B}}) \xrightarrow{\tilde{\beta}_*} H^1(\tilde{\mathcal{C}}) \xrightarrow{\tilde{\mathfrak{v}}} H^2(\tilde{\mathcal{A}}) \cdots
\end{array} .$$

Proof. The proof is called *diagram chasing* by Massey [40] (p. 184) since it "... requires practically no cleverness or ingenuity. At each stage of the proof, there is only one possible 'move'; one does not have to make any choices." The connecting homomorphism $\mathfrak{v}$ is constructed as follows. Let $c_{p-1} \in C_{p-1}$ satisfy $dc_{p-1} = 0$. Since β is assumed to be surjective, we can find $b_{p-1} \in B_{p-1}$ so that $\beta b_{p-1} = c_{p-1}$. Then

$$\beta d b_{p-1} = d\beta b_{p-1} = dc_{p-1} = 0$$

so there is a unique $a_p \in A_p$ so $\alpha a_p = db_{p-1}$. We compute $\alpha da_p = d\alpha a_p = ddb_{p-1} = 0$. Since α is injective, $da_p = 0$ so $[a_p]$ is well-defined in $H^p(\mathcal{A})$. The picture may be drawn as follows:

$$\begin{array}{ccccc}
 & & 0 & & 0 \\
 & & \downarrow & & \downarrow \\
 & & a_p & \xrightarrow{d} & da_p \\
 & & \downarrow \alpha & \circ & \downarrow \alpha \\
b_{p-1} & \xrightarrow{d} & b_p & \xrightarrow{d} & 0 \\
\downarrow \beta & \circ & \downarrow \beta & & \\
c_{p-1} & \xrightarrow{d} & 0 & & \\
\downarrow & & & & \\
0 & & & &
\end{array} .$$

Let $\tilde{b}_{p-1}$ be another possible lift, i.e., $\beta\tilde{b}_{p-1} = c_{p-1}$. Choose $\tilde{a}_p$ so that $\alpha\tilde{a}_p = d\tilde{b}_{p-1}$. As $\beta(b_{p-1} - \tilde{b}_{p-1}) = 0$, there exists a_{p-1} so $b_{p-1} - \tilde{b}_{p-1} = \alpha(a_{p-1})$. Then

$$\alpha da_{p-1} = d\alpha a_{p-1} = db_{p-1} - d\tilde{b}_{p-1} = \alpha(a_p - \tilde{a}_p) .$$

Consequently, as α is injective, $a_p - \tilde{a}_p = da_{p-1}$ so $[a_p] = [\tilde{a}_p]$ in $H^p(\mathcal{A})$. Therefore, the map $\mathfrak{v}$ which sends c_p to $[a_p]$ is a well-defined map from $\ker\{d^{\mathcal{C}}_{p-1}\}$ to $H^p(\mathcal{A})$. Next, suppose that

$c_{p-1} = dc_{p-2}$. Find b_{p-2} so $\beta b_{p-2} = c_{p-2}$. Then $\beta d b_{p-2} = d\beta b_{p-2} = dc_{p-2} = c_{p-1}$ and, consequently, db_{p-2} will do as a lift of c_{p-1}. Since $ddb_{p-2} = 0$, we have $a_p = 0$ in this instance and $\mathfrak{v}c_{p-1} = 0$. This shows $\mathfrak{v}$ is a well-defined map

$$\mathfrak{v} : H^{p-1}(\mathcal{C}) \to H^p(\mathcal{A}) .$$

We now turn to the sequence of Lemma 8.3. We begin by examining exactness at $H^p(\mathcal{B})$. Clearly, $\beta_*\alpha_*[a] = [\beta\alpha a] = 0$. Conversely, suppose $db_p = 0$ and $\beta_*[b_p] = 0$. There then exists c_{p-1} so $dc_{p-1} = \beta b_p$. Choose b_{p-1} so $\beta b_{p-1} = c_{p-1}$. Then $\beta b_p = dc_{p-1} = d\beta b_{p-1}$ and, consequently, $\beta(b_p - db_{p-1}) = 0$. Thus, there exists a_p so $b_p - db_{p-1} = \alpha a_p$. Since

$$\alpha da_p = d\alpha a_p = db_p = 0$$

and since α is injective, $da_p = 0$. Consequently, $[a_p] \in H^p(\mathcal{A})$ satisfies

$$\alpha_*[a_p] = [b_p - db_{p-1}] = [b_p] .$$

This shows

$$\ker\{\beta_*\} = \text{range}\{\alpha_*\} .$$

Next, we examine exactness at $H^{p-1}(\mathcal{C})$. Let $[c_{p-1}] \in \text{range}\{\beta_*\}$. By adjusting c_{p-1} by an element of range$\{d\}$, we may assume $c_{p-1} = \beta b_{p-1}$ where $db_{p-1} = 0$. Since $db_{p-1} = 0$, the corresponding lift $a_p = 0$ and, consequently, $\mathfrak{v}(c_{p-1}) = 0$. Therefore, $[c_{p-1}] \in \ker\{\mathfrak{v}\}$. Conversely, suppose $dc_{p-1} = 0$ and $[c_{p-1}] \in \ker\{\mathfrak{v}\}$. Choose b_{p-1} so $\beta b_{p-1} = c_{p-1}$. Express $db_{p-1} = \alpha a_p$. Since $[a_p] = \mathfrak{v}[c_{p-1}] = 0$, $a_p = da_{p-1}$. Consequently,

$$d(b_{p-1} - \alpha a_{p-1}) = db_{p-1} - \alpha a_p = 0 .$$

Consequently, $[b_{p-1} - \alpha a_{p-1}] \in H^{p-1}(\mathcal{B})$ and $\beta_*[b_{p-1} - \alpha a_{p-1}] = [c_{p-1}]$. This shows

$$\ker\{\mathfrak{v}\} = \text{range}\{\alpha^*\} .$$

Finally, we examine exactness at $H^p(\mathcal{A})$. Let $[a_p] \in \text{range}\{\mathfrak{v}\}$. Let $\alpha a_p = db_{p-1}$ for $\beta b_{p-1} = c_{p-1}$. This implies $\alpha[a_p] = 0$ so $[a_p] \in \ker\{\alpha_*\}$. Conversely, suppose $\alpha_*[a_p] = 0$. We then have $\alpha a_p = b_p$ for $b_p = db_{p-1}$. Set $c_{p-1} = \beta b_{p-1}$. Then

$$dc_{p-1} = d\beta b_{p-1} = \beta db_{p-1} = \beta b_p = \beta\alpha a_p = 0 .$$

So $[c_{p-1}] \in H^{p-1}(\mathcal{C})$ and $\mathfrak{v}[c_{p-1}] = [a_{p-1}]$. Consequently,

$$\ker\{\beta^*\} = \text{range}\{\mathfrak{v}\} .$$

This shows the sequence is exact. Clearly, $\mathfrak{v}$ is a natural map. □

8.1.5 THE 5-LEMMA. We will use the following observation repeatedly to establish the equivalence of various cohomology theories. It is called the 5-Lemma. We refer to Eilenberg and Steenrod [18] (p. 16) for details.

Lemma 8.4 Suppose given a commutative diagram of exact sequences:

$$\begin{array}{ccccccccc} A_1 & \xrightarrow{d} & A_2 & \xrightarrow{d} & A_3 & \xrightarrow{d} & A_4 & \xrightarrow{d} & A_5 \\ \downarrow \alpha_1 & \circ & \downarrow \alpha_2 & \circ & \downarrow \alpha_3 & \circ & \downarrow \alpha_4 & \circ & \downarrow \alpha_5 \\ B_1 & \xrightarrow{d} & B_2 & \xrightarrow{d} & B_3 & \xrightarrow{d} & B_4 & \xrightarrow{d} & B_5 \end{array} .$$

Assume that $\alpha_1, \alpha_2, \alpha_4$, and α_5 are isomorphisms. Then α_3 is an isomorphism.

Proof. Again the proof is a diagram chase. We shall first show that α_3 is surjective. Let $b_3 \in B_3$. Since α_4 is surjective, we may choose $a_4 \in A_4$ so $\alpha_4 a_4 = db_3$. Then

$$\alpha_5 da_4 = d\alpha_4 a_4 = ddb_3 = 0 .$$

As α_5 is injective, $da_4 = 0$. As the sequence is exact at A_4, we may choose $a_3 \in A_3$ so that $da_3 = a_4$. Let $\tilde{b}_3 = b_3 - \alpha_3 a_3$. One then has that:

$$d\tilde{b}_3 = db_3 - d\alpha_3 a_3 = db_3 - \alpha_4 da_3 = db_3 - \alpha_4 a_4 = 0 .$$

Because the sequence is exact at B_3, we may choose $b_2 \in B_2$ so that $\tilde{b}_3 = db_2$. As α_2 is surjective, we may choose $a_2 \in A_2$ so $\alpha_2 a_2 = b_2$. We have $\alpha_3 da_2 = d\alpha_2 a_2 = db_2 = b_3 - \alpha_3 a_3$ and, consequently, $b_3 = \alpha_3(a_3 + da_2)$. This shows α_3 is surjective.

We now show α_3 is injective. Suppose $\alpha_3 a_3 = 0$. Then $\alpha_4 da_3 = d\alpha_3 a_3 = 0$. Since α_4 is injective, this implies $da_3 = 0$ and hence there exists $a_2 \in A_2$ so $a_3 = da_2$ since the sequence is exact at A_3. We have

$$d\alpha_2 a_2 = \alpha_3 da_2 = \alpha_3 a_3 = 0 .$$

Since the sequence is exact at B_2, there is $b_1 \in B_1$ so $\alpha_2 a_2 = db_1$. Since α_1 is surjective, we can express $b_1 = \alpha_1 a_1$. Therefore, $\alpha_2 a_2 = db_1 = d\alpha_1 a_1 = \alpha_2 da_1$. Since α_2 is injective, $a_2 = da_1$. Consequently, $a_3 = da_2 = 0$ and α_3 is injective. □

We remark that we only used α_1 is surjective, α_2 is bijective, α_4 is bijective, and α_5 is injective so the hypotheses can be weakened slightly.

8.1.6 RING STRUCTURES IN COHOMOLOGY. Let $R = \{R_0, R_1, R_2, \dots\}$ be a collection of real (or complex) vector spaces. Denote a generic element of R_i by x_i.

1. We say that R is a *graded commutative algebra* if we have bilinear multiplication maps $\star$ from $R_j \times R_k$ to R_{j+k} which are *associative* and *skew-commutative*:

$$x_j \star (x_k \star x_\ell) = (x_j \star x_k) \star x_\ell \quad \text{and} \quad x_j \star x_k = (-1)^{jk} x_k \star x_j .$$

2. If R and $\tilde{R}$ are graded commutative algebras, we may define $S := R \otimes \tilde{R}$ by setting

$$S_i = \oplus_{p+q=i} R_p \otimes \tilde{R}_q \quad \text{and} \quad (x_i \otimes \tilde{x}_j) \star (x_k \otimes \tilde{x}_\ell) := (-1)^{jk} (x_i \star x_k) \otimes (\tilde{x}_j \star \tilde{x}_\ell) .$$

3. We say a graded commutative algebra R is *connected and unital* if

$$R_0 = \mathbb{R} \cdot 1 \quad \text{and} \quad 1 \star x_j = x_j \star 1 \quad \text{for any} \quad x_j \in R_j .$$

If R and $\tilde{R}$ are connected and unital, then $R \otimes \tilde{R}$ is connected and unital where

$$1_{R \otimes \tilde{R}} := 1_R \otimes 1_{\tilde{R}} .$$

4. If R and $\tilde{R}$ are two graded, commutative, connected, and unital algebras, then we say that a collection $T = \{T_i\}$ of linear maps $T_i : R_i \to \tilde{R}_i$ is an *algebra morphism* if

$$T_{j+k}(x_j \star x_k) = T_j(x_j) \star T_k(x_k) \quad \text{and} \quad T_0(1_R) = 1_{\tilde{R}} .$$

5. We say that C is a graded commutative cochain complex if C is a graded commutative algebra, and if the differential d satisfies $d(x_p \star x_q) = dx_p \star x_q + (-1)^p x_p \star dx_q$. A morphism in this context is a graded commutative algebra morphism which commutes with the differential.

Lemma 8.5 If C is a graded commutative algebra which is a cochain complex, then the map $[x] \star [y] := [x \star y]$ gives the cohomology $H^*(C)$ the structure of a graded commutative algebra. If $T : C \to \tilde{C}$ is a morphism of graded commutative algebra cochain complexes, then the map T_* in cohomology is a morphism of graded commutative algebras.

Proof. If $dx_p = 0$ and if $dy_q = 0$, then $d(x_p \star y_q) = dx_p \star y_q + (-1)^p x_p \star dy_q = 0$. Consequently, $[x_p \star y_q]$ is a cohomology class. If $x_p = dz_{p-1}$, then

$$d(z_{p-1} \star y_q) = dz_{p-1} \star y_q + (-1)^{p-1} z_{p-1} \star dy_q = x_p \star y_q + 0 .$$

Therefore, $[x_p] = 0$ in cohomology implies $[x_p \star y_q] = 0$ in cohomology. A similar argument shows that if $[y_q] = 0$ in cohomology, then $[x_p \star y_q] = 0$ in cohomology. Consequently, the map

$$([x_p], [y_q]) \to [x_p \star y_q]$$

is well-defined and gives a bilinear map from the tensor product $H^p(C) \otimes H^q(C)$ to $H^{p+q}(C)$. The identity $x_p \star y_q = (-1)^{pq} y_q \star x_p$ implies the multiplication is skew-commutative. A similar argument shows that the multiplication is associative. The assertion about morphisms follows similarly. □

8.1.7 THE MAYER–VIETORIS SEQUENCE. We complete the proof of Theorem 5.2 by showing the existence of the Mayer–Vietoris sequence in de Rham cohomology and by showing de Rham cohomology is a homotopy functor; this verifies that de Rham cohomology satisfies the Eilenberg–Steenrod axioms [17, 18]. We restate these two properties for ease of reference.

Theorem 8.6

1. *If $\mathcal{O}_i$ are open subsets of M with $M = \mathcal{O}_1 \cup \mathcal{O}_2$, then there is a natural long exact sequence (called the Mayer–Vietoris sequence [41, 59]):*

$$\cdots \to H^{p-1}_{\mathrm{dR}}(M) \stackrel{i_1^* \oplus i_2^*}{\longrightarrow} H^{p-1}(\mathcal{O}_1) \oplus H^{p-1}(\mathcal{O}_2) \stackrel{j_1^* - j_2^*}{\longrightarrow} H^{p-1}_{\mathrm{dR}}(\mathcal{O}_1 \cap \mathcal{O}_2) \stackrel{\mathfrak{v}}{\longrightarrow} H^p_{\mathrm{dR}}(M) \to \cdots$$

where we take the natural inclusions:

$$i_1 : \mathcal{O}_1 \to M, \quad i_2 : \mathcal{O}_2 \to M, \qquad j_1 : \mathcal{O}_1 \cap \mathcal{O}_2 \to \mathcal{O}_1, \quad j_2 : \mathcal{O}_1 \cap \mathcal{O}_2 \to \mathcal{O}_2 .$$

The map $\mathfrak{v}$ is called the connecting homomorphism; if $f : N \to M$ and if $\mathcal{U}_i := f^{-1}\mathcal{O}_i$ is the associated open cover of N, then $\mathfrak{v}_N f^* = f^* \mathfrak{v}_M$, i.e., $\mathfrak{v}$ is natural in this category.

2. *If $f_i : M \to N$ are homotopic smooth maps, then $f_0^* = f_1^* : H^p_{\mathrm{dR}}(N) \to H^p_{\mathrm{dR}}(M)$.*

Proof. To prove Assertion 1, we let $\mathcal{O}_1$ and $\mathcal{O}_2$ be open subsets of M with $M = \mathcal{O}_1 \cup \mathcal{O}_2$. Let $\mathcal{D}(\cdot)$ denote the de Rham complex

$$0 \to C^\infty(\Lambda^0(\cdot)) \stackrel{d_0}{\longrightarrow} C^\infty(\Lambda^1(\cdot)) \stackrel{d_1}{\longrightarrow} C^\infty(\Lambda^2(\cdot)) \to \cdots .$$

We consider the sequence of cochain complexes:

$$0 \to \mathcal{D}(M) \stackrel{i_1^* \oplus i_2^*}{\longrightarrow} \mathcal{D}(\mathcal{O}_1) \oplus \mathcal{D}(\mathcal{O}_2) \stackrel{j_1^* - j_2^*}{\longrightarrow} \mathcal{D}(\mathcal{O}_1 \cap \mathcal{O}_2) \to 0 .$$

If we can show this is a short exact sequence of cochain complexes, then Assertion 1 will follow from Lemma 8.3. Let $\Theta_p \in C^\infty(\Lambda^p(M))$. If $i_i \Theta_p = 0$, then Θ_p vanishes on $\mathcal{O}_i$. If Θ_p vanishes on $\mathcal{O}_1$ and on $\mathcal{O}_2$, then Θ_p vanishes on M. Consequently, $(i_1^* \oplus i_2^*)\Theta_p = 0$ if and only if $\Theta_p = 0$. This verifies exactness of the sequence at $\mathcal{D}(M)$. We have:

$$(j_1^* - j_2^*) \circ (i_1^* \oplus i_2^*) = j_1^* i_1^* - j_2^* i_2^* = (i_1 j_1)^* - (i_2 j_2)^* = 0$$

since $i_1 j_1 = i_2 j_2$ is just the inclusion of $\mathcal{O}_1 \cap \mathcal{O}_2$ in M. Conversely, suppose $j_1^*\theta_1 - j_2^*\theta_2 = 0$. This means the restriction of θ_1 agrees with the restriction of θ_2 to $\mathcal{O}_1 \cap \mathcal{O}_2$. Consequently, we may define θ to be θ_1 on $\mathcal{O}_1$ and θ_2 on $\mathcal{O}_2$. This defines a p-form on M which restricts to θ_i on $\mathcal{O}_i$. This verifies exactness of the sequence at the middle term $\mathcal{D}(\mathcal{O}_1) \oplus \mathcal{D}(\mathcal{O}_2)$. Finally, we must verify that $j_1^* - j_2^*$ is surjective. We use an argument which we will employ subsequently when considering sheaf cohomology. Let $\{\phi_1, \phi_2\}$ be a partition of unity subordinate to the cover $\{\mathcal{O}_1, \mathcal{O}_2\}$ of M. Let $\theta \in C^\infty(\Lambda^p(\mathcal{O}_1 \cap \mathcal{O}_2))$. Define:

$$\theta_1(x_1) := \left\{ \begin{array}{ll} \phi_2(x_1)\theta(x_1) & \text{if } x_1 \in \mathcal{O}_1 \cap \mathcal{O}_2 \\ 0 & \text{if } x_1 \in \mathcal{O}_1 \cap \mathcal{O}_2^c \end{array} \right\},$$

$$\theta_2(x_2) := \left\{ \begin{array}{ll} \phi_1(x_2)\theta(x_2) & \text{if } x_2 \in \mathcal{O}_1 \cap \mathcal{O}_2 \\ 0 & \text{if } x_2 \in \mathcal{O}_1^c \cap \mathcal{O}_2 \end{array} \right\}.$$

We must verify θ_i is smooth on $\mathcal{O}_i$. Since $\theta_1 = \phi_2\theta$ on $\mathcal{O}_1 \cap \mathcal{O}_2$, θ_1 is smooth on $\mathcal{O}_1 \cap \mathcal{O}_2$. Suppose $x \in \mathcal{O}_1 \cap \mathcal{O}_2^c$. The support of ϕ_2 is contained in $\mathcal{O}_2$. Let $\mathcal{U} := \mathcal{O}_1 \cap \text{support}\{\phi_2\}^c$. Then $x \in \mathcal{U}$ and $\phi_2 = 0$ on $\mathcal{U}$. Consequently, θ_1 is smooth on $\mathcal{U}$ as well. Since $\{\mathcal{O}_1 \cap \mathcal{O}_2, \mathcal{U}\}$ forms an open cover of $\mathcal{O}_1$, $\theta_1 \in C^\infty(\Lambda^p(\mathcal{O}_1))$; similarly one has that $\theta_2 \in C^\infty(\mathcal{O}_2)$. Since

$$j_1^*\theta_1 - j_2^*(-\theta_2) = (\phi_2 + \phi_1)\theta = \theta$$

the sequence is exact at $\mathcal{D}(\mathcal{O}_1 \cap \mathcal{O}_2)$. This establishes Assertion 1.

To establish Assertion 2, we shall construct a suitable chain homotopy and apply Lemma 8.1. Let $X \in C^\infty(TM)$ and let $\theta \in C^\infty(\Lambda^p M)$. Define $\text{int}(X)\theta \in C^\infty(\Lambda^{p-1}M)$ by:

$$\{\text{int}(X)\theta\}(X_2, \dots, X_{p-1}) = \theta(X, X_1, \dots, X_{p-1}). \tag{8.1.b}$$

For example, if $\vec{x} = (x^1, \dots, x^m)$ is a system of local coordinates, then

$$\text{int}(\partial_{x^1})\{dx^{i_1} \wedge \dots \wedge dx^{i_p}\} = \left\{\begin{array}{ll} 0 & \text{if } i_1 > 1 \\ dx^{i_2} \wedge \dots \wedge dx^{i_p} & \text{if } i_1 = 1 \end{array}\right\}.$$

This is also often called the *hook product* and is the dual of the interior product $\text{int}(\xi)$ for $\xi \in T^*M$ defined in Section 5.2. Let $F : M \times [0,1] \to N$ and let $\theta \in C^\infty(\Lambda^p N)$. Define

$$\Xi\theta(x) := \int_0^1 \{\text{int}(\partial_t)F^*\theta\}(x;t)dt \in C^\infty(\Lambda^{p-1}M).$$

This is invariantly defined. Introduce local coordinates $(x^1, \dots, x^m; t)$ on $M \times [0,1]$. Expand

$$F^*\theta = \sum_{|I|=p} \Theta_I(x;t)dx^I + \sum_{|J|=p-1} \tilde{\Theta}_J(x;t)dt \wedge dx^J.$$

We show that Ξ is the desired chain homotopy by computing:

$$(f_1^* - f_0^*)\theta = \sum_{|I|=p} \{\Theta_I(x;1) - \Theta_I(x;0)\}dx^I,$$

$$\Xi d\theta = \Xi\left\{\sum_{I,i} \partial_{x^i}\Theta_I dx^i \wedge dx^I + \sum_I \partial_t\Theta_I dt \wedge dx^I - \sum_{J,i} \partial_{x^i}\tilde{\Theta}_J dt \wedge dx^i \wedge dx^J\right\}$$

$$= 0 + (f_1^* - f_0^*)\sum_I \theta_I dx^I - \left\{\int_0^1 \partial_{x^i}\tilde{\Theta}_J(x;t)dt\right\} dx^i \wedge dx^J,$$

$$d\Xi\theta = \sum_{i,J}\left\{\int_0^1 \partial_{x^i}\tilde{\Theta}_J(x;t)dt\right\}dx^i \wedge dx^J.$$

It is now immediate that $d\Xi + \Xi d = f_1^* - f_0^*$ so Assertion 2 follows from Lemma 8.1. □

8.2 SIMPLICIAL COHOMOLOGY

We begin by defining the notion of a finite simplicial complex K and the associated realization $|K|$; K is a combinatorial object and $|K|$ is a compact metric space. Although much of what we will say works for infinite simplicial complexes, more care needs to be taken with the topology involved and the arguments that we shall give by induction either fail in the more general setting, or need to be reformulated.

8.2.1 SIMPLICIAL COMPLEXES. A *finite simplicial complex* consists of finite set of vertices $V = \{v_0, \dots, v_\ell\}$ together with a collection K of subsets of V so that the empty set belongs to K, the singleton set $\{v\}$ for any $v \in V$ belongs to K, and if $A \in K$ and if B is any subset of A, then $B \in K$. The vertex set $V = V(K)$ is the union of the singleton sets in K so it need not be specified separately.

Let K be a finite simplicial complex. Let $C_0^{\text{PL}}(K)$ be the finite-dimensional $\mathbb{R}$ vector space with basis $V(K)$. We introduce a positive definite inner product on $C_0^{\text{PL}}(K)$ by requiring that the vertices form an orthonormal basis. This makes $C_0^{\text{PL}}(K)$ into an inner product space and defines a topology on $C_0^{\text{PL}}(K)$. If $x \in C_0^{\text{PL}}(K)$, expand $x = \sum_{v\in V(K)} x(v)\cdot v$ where $x(v) \in \mathbb{R}$ are the coefficient functions for $v \in V$. Let $\text{support}(x) = \{v : x(v) \neq 0\}$. Define the *realization* of K by setting:

$$|K| := \left\{ x \in C_0^{\text{PL}}(K) : x(v) \geq 0 \quad \forall\ v, \quad \sum_{v\in V(K)} x(v) = 1, \quad \text{support}(x) \in K \right\}.$$

The line segment $I(v_{i_1}, v_{i_2})$ with vertices $\{v_{i_1}, v_{i_2}\}$ is parameterized by

$$t^1 v_{i_1} + t^2 v_{i_2} \quad \text{for} \quad 0 \leq t^1, \quad 0 \leq t^2, \quad t^1 + t^2 = 1\,.$$

Note that $I(v_{i_1}, v_{i_2}) \subset |K|$ if and only if $\{v_{i_1}, v_{i_2}\} \in K$. The triangle $T(v_{i_1}, v_{i_2}, v_{i_3})$ with vertices at $\{v_{i_1}, v_{i_2}, v_{i_3}\}$ is parameterized by

$$t^1 v_{i_1} + t^2 v_{i_2} + t^3 v_{i_3} \quad \text{for} \quad 0 \leq t^1, \quad 0 \leq t^2, \quad 0 \leq t^3, \quad \text{and} \quad t^1 + t^2 + t^3 = 1\,.$$

Note that $T(v_{i_1}, v_{i_2}, v_{i_3}) \subset |K|$ if and only if $\{v_{i_1}, v_{i_2}, v_{i_3}\} \in K$. Consequently, we may think of $|K|$ as a children's toy where we glue in edges, faces, etc. according to the combinatorial recipe provided by K. We refer to H. Whitney [62, 63] for the proof of the following result as it is a bit beyond the scope of this book.

Theorem 8.7 *Let M be a compact manifold. Then there exists a finite simplicial complex K so that M is homeomorphic to $|K|$.*

Let $|V| = m + 1$. If $K = 2^V$ is the collection of all subsets of V, then $|K|$ is homeomorphic to the unit disk D^m in $\mathbb{R}^m$. If $m = 1$, then $|K|$ is the interval; if $m = 2$, then $|K|$ is the solid triangle; if $m = 3$, then $|K|$ is the solid tetrahedron. Let L be the collection of all proper subsets of V. Then $|L|$ is homeomorphic to the boundary of D^m, i.e., to the sphere S^{m-1}.

We now discuss simplicial cohomology; this is a purely combinatorial object. Let K be a finite simplicial complex. We defined $C_0^{\text{PL}}(K)$ to be the real vector space with basis the vertices of K. If $A = \{v_{i_0}, \dots, v_{i_q}\}$ is a q-simplex of K, let

$$v_A := v_{i_0} \wedge \dots \wedge v_{i_q} \in \Lambda^{q+1}(C_0^{\text{PL}}(K)),$$
$$C_q^{\text{PL}}(K) := \operatorname{span}_{A \in K, |A|=q+1}\{v_A\} \subset \Lambda^{q+1}(C_0^{\text{PL}}(K))\,.$$

The $\{v_A\}$ for $i_0 < \dots < i_q$ form an orthonormal basis for $\Lambda^{q+1}(C_0^{\text{PL}}(K))$ with the induced inner product. Let

$$\delta_{\text{PL}} := \sum_{v \in V} \operatorname{int}(v), \qquad \delta_{\text{PL}}(v_A) := \sum_{j=0}^{q} (-1)^j v_{i_0} \wedge \dots \wedge v_{i_{j-1}} \wedge v_{i_{j+1}} \wedge \dots \wedge v_{i_q}\,.$$

We dualize and set

$$C_{\text{PL}}^q(K) := \operatorname{Hom}(C_q^{\text{PL}}(K), \mathbb{R}) \quad \text{and} \quad d_{\text{PL}} = \delta_{\text{PL}}^* : C_{\text{PL}}^{q-1}(K) \to C_{\text{PL}}^q(K)\,.$$

We have chosen to use the notation δ_{PL} for the boundary operator so that the coboundary operator on the associated cochain complex will be d_{PL}; this is somewhat different notation than is usually employed.

Lemma 8.8 If K is a simplicial complex, then $\delta_{\text{PL}}^2 = 0$ and $d_{\text{PL}}^2 = 0$.

Proof. By Lemma 5.8, $\operatorname{int}(v)\operatorname{int}(w) + \operatorname{int}(w)\operatorname{int}(v) = 0$. We will show that $\delta_{\text{PL}}^2 = 0$; we then have dually $d_{\text{PL}}^2 = (\delta_{\text{PL}}^*)^2 = (\delta_{\text{PL}}^2)^* = 0$. We compute:

$$\delta_{\text{PL}}^2(A) = \sum_{v,w \in V} \operatorname{int}(v)\operatorname{int}(w) v_A = -\sum_{v,w \in V} \operatorname{int}(w)\operatorname{int}(v) v_A = -\delta_{\text{PL}}^2(v_A)\,. \qquad \square$$

Because $d_{\text{PL}}^2 = 0$, $(C_{\text{PL}}^*(K), d_{\text{PL}})$ forms a cochain complex and we define the *PL cohomology* of K by setting:

$$H_{\text{PL}}^q(K) := \frac{\ker\{d_{\text{PL}} : C_{\text{PL}}^q(K) \to C_{\text{PL}}^{q+1}(K)\}}{\operatorname{range}\{d_{\text{PL}} : C_{\text{PL}}^{q-1}(K) \to C_{\text{PL}}^q(K)\}}\,.$$

8.2.2 SIMPLICIAL MAPS. Let K and L be simplicial complexes. A *simplicial map* is a map of the vertex sets $f : V_K \to V_L$ so that if $A \in K$, then $f(A) \in L$. Let

$$f_*(v_{i_0} \wedge \dots \wedge v_{i_p}) := f(v_{i_0}) \wedge \dots \wedge f(v_{i_p})\,.$$

Since $\delta_L f_* = f_* \delta_K$, $d_K f^* = f^* d_L$. Consequently, we have a map of cochain complexes:

$$f^* : (C_*^{\text{PL}}(L), d_L) \to (C_{\text{PL}}^*(K), d_K)\,.$$

Lemma 8.1 yields corresponding maps in PL cohomology. We use the inner product on $\Lambda^q(C_0^{\mathrm{PL}}(K))$ to identify $C^q_{\mathrm{PL}}(K)$ with $C_q^{\mathrm{PL}}(K)$; under this identification d_{PL} is the linear adjoint of δ_{PL}. We define the PL Laplacian by setting

$$\Delta_{\mathrm{PL}} := d_{\mathrm{PL}}\delta_{\mathrm{PL}} + \delta_{\mathrm{PL}}d_{\mathrm{PL}}\,.$$

We say K_1 is a *subsimplicial complex* of K if $K_1 \subset K$ is a simplicial complex in its own right where the vertex set $V(K_1)$ are just the singletons of K_1, i.e., $V(K_1) = K_1 \cap V(K)$. If K_1 and K_2 are subsimplicial complexes of K, then $K_1 \cap K_2$ is a subsimplicial complex of K with corresponding vertex sets $V(K_1) \cap V(K_2)$. Finally, let $\chi_{\mathrm{PL}}(K) = \sum_{\emptyset \neq A \in K}(-1)^{|A|}$ be number of vertices minus number of edges plus number of triangles etc.; this is the *combinatorial Euler characteristic*. One has the following result.

Theorem 8.9 *Let K be a finite simplicial complex.*

1. *If $f : K \to L$ is a simplicial map, then f^* induces a natural map in cohomology from $H^p_{PL}(L)$ to $H^p_{PL}(K)$ and makes $H^p_{PL}(\cdot)$ into a contravariant functor.*
2. *Let K_1 and K_2 be subsimplicial complexes of K so $K = K_1 \cup K_2$. There is a natural long exact sequence (Mayer–Vietoris)*
$$\cdots \to H^{p-1}_{PL}(K) \xrightarrow{i_1^* \oplus i_2^*} H^{p-1}_{PL}(K_1) \oplus H^{p-1}_{PL}(K_2) \xrightarrow{j_1^* - j_2^*} H^{p-1}_{PL}(K_1 \cap K_2) \xrightarrow{\mathfrak{d}} H^p_{PL}(K) \to \cdots$$
where $i_1 : K_1 \to K$, $i_2 : K_2 \to K$, $j_1 : K_1 \cap K_2 \to K_1$, and $j_2 : K_1 \cap K_2 \to K_2$ are the natural inclusions. The map $\mathfrak{d}$ is called the connecting homomorphism.
3. *If K is the disjoint union of two simplicial complexes K_1 and K_2, then*
$$H^q_{PL}(K) = H^q_{PL}(K_1) \oplus H^q_{PL}(K_2)\,.$$
4. *$H^q_{PL}(K) = \ker\{\Delta^q_{PL}\}$ and $\chi_{PL}(K) = \sum_q (-1)^q \dim\{H^q_{PL}(K)\}$.*
5. *If $K = 2^S$ where S is finite, then* $H^q_{PL}(K) = \left\{\begin{array}{ll} \mathbb{R} & \text{if } q = 0 \\ 0 & \text{if } q > 0 \end{array}\right\}$.

Proof. Assertion 1 is immediate from the definition. We may use Lemma 8.3 to establish Assertion 2 since the short exact sequence of chain complexes

$$0 \to C_*^{\mathrm{PL}}(K_1 \cap K_2) \to C_*^{\mathrm{PL}}(K_1) \oplus C_*^{\mathrm{PL}}(K_2) \to C_*^{\mathrm{PL}}(K) \to 0$$

gives rise dually to a corresponding short exact sequence of cochain complexes:

$$0 \to C^*_{\mathrm{PL}}(K) \to C^*_{\mathrm{PL}}(K_1) \oplus C^*_{\mathrm{PL}}(K_2) \to C^*_{\mathrm{PL}}(K_1 \cap K_2) \to 0\,.$$

Assertion 3 is again immediate from the definition as the cochain complexes decouple. The first identity of Assertion 4 follows immediately from Lemma 8.2 and the remaining identity follows from Lemma 8.2 since $\chi_{\mathrm{PL}}(K) = \sum_i (-1)^i \dim\{C^i_{\mathrm{PL}}(K)\}$. We argue as follows to prove Assertion 5. Let $A \in K$ with $|A| \geq 2$. Then

$$\delta_{\mathrm{PL}}(v_A) = \sum_{v \in V} \mathrm{int}(v) v_A \quad \text{and} \quad d_{\mathrm{PL}}(v_A) = \sum_{v \in V : \{a, A\} \in K} \mathrm{ext}(v) v_A \,.$$

The restriction that $\{a, A\} \in K$ is, of course, necessary to ensure that $\mathrm{ext}(v) v_A \in C^{q+1}_{\mathrm{PL}}(K)$. But if $K = 2^S$, the restriction is unnecessary and Lemma 5.8 yields

$$\Delta^q_{\mathrm{PL}} v_A = \sum_{v, w \in S} \{\mathrm{ext}(v)\,\mathrm{int}(w) + \mathrm{int}(w)\,\mathrm{ext}(v)\} v_A = |S| v_A \,.$$

Consequently, $\ker\{\Delta^q_{\mathrm{PL}}\} = \{0\}$ and $H^q_{\mathrm{PL}}(K) = 0$ if $q > 0$. We use Assertion 4 to see that:

$$\chi_{\mathrm{PL}}(K) = \sum_q (-1)^1 \dim\{H^q_{\mathrm{PL}}(K)\} = \dim\{H^0_{\mathrm{PL}}(K)\} \,.$$

Let $|S| = m$. We complete the proof of Assertion 5 by nothing that:

$$\chi_{\mathrm{PL}}(K) = m - \binom{m}{2} + \binom{m}{3} + \cdots + (-1)^m \binom{m}{m} = 1 \,. \qquad \square$$

8.3 SINGULAR COHOMOLOGY

If K and L are simplicial complexes, and if $f : K \to L$ is a simplicial map, there is a natural induced map $|f| : |K| \to |L|$ defined by setting $|f|(t^i v_i) = t^i f(v_i)$. Let $S_q = \{e_0, \ldots, e_q\}$, let $K_q = 2^{S_q}$, and let $\sigma_q := |S_q|$ be the standard q-simplex. Let τ_j be the face map from S_{q-1} to S_q defined by:

$$\tau_j(e_i) := e_i \quad \text{if} \quad i < j \quad \text{and} \quad \tau_j(e_i) = e_{i+1} \quad \text{if} \quad i \geq j \,.$$

The map τ_j is a simplicial map and $|\tau_j| : \sigma_{q-1} \to \sigma_q$. In other words, we put the $q - 1$ simplex S_{q-1} into the q simplex S_q by laying it along the $q - 1$ simplex obtained by deleting the ith vertex. Consequently, for example, the boundary operator δ_{PL} is given by

$$\delta_{\mathrm{PL}}(S_q) = \sum_{i=0}^{q} (-1)^i \tau_i(S_{q-1}) \,.$$

If $q = 1$, then $\tau_0(0) = e_1$ and $\tau_1(0) = e_0$ so the boundary of the 1-simplex (e_0, e_1) is simply $(e_1) - (e_0)$. In terms of Stokes' Theorem,

$$\mathrm{bd}(S_q) = |\delta_{\mathrm{PL}}|(|S_{q-1}|) = \sum_{i=0}^{q} (-1)^j |\tau_j|(|S_{q-1}|)$$

where we keep track of the orientation corresponding to the ordering of the vertices. Let X be a topological space. Let C^{TP}_q be the $\mathbb{R}$ vector space with basis the continuous maps $f_q : \sigma_q \to X$.

The boundary map δ_{TP} from $C_q^{\mathrm{TP}}(X)$ to $C_{q-1}^{\mathrm{TP}}(X)$ is defined by

$$\delta_{\mathrm{TP}}(f_q) = \sum_{j=0}^{q}(-1)^j f_q \circ |\tau_j|\,.$$

An easy calculation along the lines of those performed previously shows that $\delta_{\mathrm{TP}}^2 = 0$. Let $C_{\mathrm{TP}}^q(X) := \mathrm{Hom}(C_q^{\mathrm{TP}}, \mathbb{R})$ and let $\langle\cdot,\cdot\rangle$ be the pairing between $C_{\mathrm{TP}}^q(X)$ and $C_q^{\mathrm{TP}}(X)$; if ω belongs to $C_{\mathrm{TP}}^q(X)$ and σ belongs to $C_q^{\mathrm{TP}}(X)$, then $\langle\omega,\sigma\rangle \in \mathbb{R}$. The derivative d_{TP} from $C_{\mathrm{TP}}^{q-1}(X)$ to $C_{\mathrm{TP}}^q(X)$ is defined dually to be d_{TP}^*; if $\omega \in C_{\mathrm{TP}}^q(X)$ and if $\sigma \in C_{q+1}^{\mathrm{TP}}(X)$, we have

$$\langle d\omega, \sigma\rangle = \langle\omega, \delta\sigma\rangle\,.$$

We have $\delta_{\mathrm{TP}}^2 = 0$ so dually $d_{\mathrm{TP}}^2 = 0$. We define the singular cohomology groups to be:

$$H_{\mathrm{TP}}^q(X) := \frac{\ker\{d_{\mathrm{TP}} : C_{\mathrm{TP}}^q(X) \to C_{\mathrm{TP}}^{q+1}(X)\}}{\mathrm{range}\{d_{\mathrm{TP}} : C_{\mathrm{TP}}^{q-1}(X) \to C_{\mathrm{TP}}^q(X)\}}\,.$$

If $f : X \to Y$ is a continuous map, we set $f_*\sigma = f \circ \sigma$ and extend linearly to define a chain map f_*, which is called *pushforward* from $(C_q^{\mathrm{TP}}(X), \delta_X)$ to $(C_q^{\mathrm{TP}}(Y), \delta_Y)$; the dual cochain map f^* from $(C_{\mathrm{TP}}^q(Y), d_Y)$ to $(C_{\mathrm{TP}}^q(X), d_X)$ is called *pullback*. Theorem 5.2 generalizes to this setting; $H_{\mathrm{TP}}^q(\cdot)$ satisfies the Eilenberg–Steenrod axioms. We shall omit the proof and instead refer to Eilenberg and Steenrod [17, 18] and to Spanier [56].

Theorem 8.10 *Let X and Y be topological spaces.*

1. *If $f : X \to Y$, then $f^* : H_{\mathrm{TP}}^p(Y) \to H_{\mathrm{TP}}^p(X)$; $\mathrm{Id}^* = \mathrm{Id}$ and $(f \circ g)^* = g^* \circ f^*$.*
2. *If X consists of a single point, then $H_{\mathrm{TP}}^0(X) = \mathbb{R}$ if $p = 0$ and 0 if $p > 0$.*
3. *If $\mathcal{O}_i$ are open sets with $X = \mathcal{O}_1 \cup \mathcal{O}_2$, then there is a natural long exact sequence (called the Mayer–Vietoris sequence [41, 59]):*
$$\cdots \to H_{\mathrm{TP}}^{p-1}(X) \xrightarrow{i_1^*\oplus i_2^*} H_{\mathrm{TP}}^{p-1}(\mathcal{O}_1) \oplus H_{\mathrm{TP}}^{p-1}(\mathcal{O}_2) \xrightarrow{j_1^*-j_2^*} H_{\mathrm{TP}}^{p-1}(\mathcal{O}_1 \cap \mathcal{O}_2) \xrightarrow{\mathfrak{d}} H_{\mathrm{TP}}^p(X) \to \cdots$$
where $i_1 : \mathcal{O}_1 \to X$, $i_2 : \mathcal{O}_2 \to X$, $j_1 : \mathcal{O}_1 \cap \mathcal{O}_2 \to \mathcal{O}_1$, and $j_2 : \mathcal{O}_1 \cap \mathcal{O}_2 \to \mathcal{O}_2$ are the natural inclusions. The map $\mathfrak{d}$ in the Mayer–Vietoris sequence is called the connecting homomorphism.
4. *If $f_i : X \to Y$ are homotopic maps, then $f_0^* = f_1^* : H_{\mathrm{TP}}^p(Y) \to H_{\mathrm{TP}}^p(X)$.*

Let K be a finite simplicial complex. Order the vertices of K. Let $A = \{v_{i_0}, \dots, v_{i_q}\}$ be a q-simplex of K where $v_{i_0} < \cdots < v_{i_q}$. Define a continuous map σ_A from the standard q-simplex with vertices $\{e_0, \dots, e_q\}$ to $|K|$ by setting:

$$\sigma_A(t_0e_0 + \cdots + t_qe_q) := t_0v_{i_0} + \cdots + t_qv_{i_q}\,.$$

The map $A \to \sigma_A$ defines a linear map σ_K from $C_q^{\mathrm{PL}}(K)$ to $C_q^{\mathrm{TP}}(|K|)$ which is a chain map. Dually, we have $\sigma_K^* : C_{\mathrm{TP}}^q(|K|) \to C_{\mathrm{PL}}^q(K)$ is a cochain map.

Theorem 8.11 *Let K be a finite simplicial complex. The cochain map σ_K^* from $C_{\mathrm{TP}}^*(|K|, d)$ to $C_{PL}^*(K, d)$ defines a natural isomorphism in cohomology from $H_{\mathrm{TP}}^q(|K|)$ to $H_{PL}^q(K)$.*

Proof. Suppose that K has only one vertex v_0 so $K = \{\emptyset, v_0\}$ consists of the empty set and a singleton set. We then have that $H_{\mathrm{PL}}^q(K) = H_{\mathrm{TP}}^q(|K|) = 0$ for $q > 0$ so σ_K^* is trivially an isomorphism in these degrees. Furthermore, $C_{\mathrm{PL}}^0(K) = C_{\mathrm{TP}}^0(|K|) = \mathbb{R}$, σ_K is the identity map, and $d_{\mathrm{PL}} = d_{\mathrm{TP}} = 0$. Therefore, in this instance, Theorem 8.11 follows for this special case. If there are no simplices of higher dimension, then K consists solely of the empty set and singleton sets so $K = \{\emptyset, V\}$ where $V = V(K)$ is the set of vertices. Let $K_v = \{\emptyset, \{v\}\}$ for $v \in V$. The problem decouples and we have

$$H_{\mathrm{PL}}^*(K) = \oplus_{v \in V} H_{\mathrm{PL}}^*(K_v), \quad H_{\mathrm{TP}}^*(K) = \oplus_{v \in V} H_{\mathrm{TP}}^*(K_v), \quad \tau_K = \oplus_{v \in V} \tau_{K_v}$$

so Theorem 8.11 follows from the case $K = K_v$ for a single vertex which was considered above. We therefore may proceed by induction on the cardinality $|K|$ and assume K contains a q-simplex for $q > 0$. Choose $S \in K$ so $|S| > 0$ is maximal. Let

$$A = 2^S, \quad B = K - S, \quad C = A - S = A \cap B .$$

Let $\mathfrak{A} := |A|$, $\mathfrak{B} := |B|$, $\mathfrak{C} := |C|$, and $\mathfrak{K} := |K|$. We consider the commutative diagram defined by the maps σ_A, σ_B, σ_C, and σ_K:

$$\begin{array}{ccccccccc}
H_{\mathrm{TP}}^{q-1}(\mathfrak{A}) \oplus H_{\mathrm{TP}}^{q-1}(\mathfrak{B}) & \rightarrow & H_{\mathrm{TP}}^{q-1}(\mathfrak{C}) & \rightarrow & H^q(\mathfrak{K}) & \rightarrow & H_{\mathrm{TP}}^q(\mathfrak{A}) \oplus H_{\mathrm{TP}}^q(\mathfrak{B}) & \rightarrow & H_{\mathrm{TP}}^q(\mathfrak{C}) \\
\sigma_A^* \oplus \sigma_B^* \downarrow\approx & \circ & \sigma_C^* \downarrow\approx & \circ & \sigma_K^* \downarrow & \circ & \sigma_A^* \oplus \sigma_B^* \approx\downarrow & \circ & \sigma_C^* \downarrow\approx \\
H_{\mathrm{PL}}^{q-1}(A) \oplus H_{\mathrm{PL}}^{q-1}(B) & \rightarrow & H_{\mathrm{PL}}^{q-1}(C) & \rightarrow & H^q(K) & \rightarrow & H_{\mathrm{PL}}^q(A) \oplus H_{\mathrm{PL}}^q(B) & \rightarrow & H_{\mathrm{PL}}^q(C) .
\end{array}$$

The bottom row is part of a long exact sequence. If we could show that the top row was part of a long exact sequence, we could then use the 5-Lemma (see Lemma 8.4) to show that σ_K^* was an isomorphism and complete the proof of Theorem 8.11.

Let $\mathcal{A} := \{x \in \mathfrak{K} : \operatorname{support}(x) \cap S \neq \emptyset\}$, $\mathcal{B} := \mathfrak{K} - \{\mathrm{pt}\}$ where $\{\mathrm{pt}\}$ is a point in the interior of $\mathfrak{A}$, and $\mathcal{C} := \mathcal{A} \cap \mathcal{B}$ be small open neighborhoods of $\mathfrak{A}$, $\mathfrak{B}$, and $\mathfrak{C}$ which deformation retract to $\mathfrak{A}$, $\mathfrak{B}$, and $\mathfrak{C}$. Let i_A, i_B, i_C, and i_K be the natural inclusions. We could then consider the diagram:

$$\begin{array}{ccccccccc}
H_{\mathrm{TP}}^{q-1}(\mathcal{A}) \oplus H_{\mathrm{TP}}^{q-1}(\mathcal{B}) & \rightarrow & H_{\mathrm{TP}}^{q-1}(\mathcal{C}) & \rightarrow & H^q(\mathfrak{K}) & \rightarrow & H_{\mathrm{TP}}^q(\mathcal{A}) \oplus H_{\mathrm{TP}}^q(\mathcal{B}) & \rightarrow & H_{\mathrm{TP}}^q(\mathcal{C}) \\
\downarrow\approx i_A^* \oplus i_B^* & \circ & \downarrow\approx i_C^* & \circ & \downarrow\approx i_K^* & \circ & \downarrow\approx i_A^* \oplus i_B^* & \circ & \downarrow\approx i_C^* \\
H_{\mathrm{TP}}^{q-1}(\mathfrak{A}) \oplus H_{\mathrm{TP}}^{q-1}(\mathfrak{B}) & \rightarrow & H_{\mathrm{TP}}^{q-1}(\mathfrak{C}) & \rightarrow & H^q(\mathfrak{K}) & \rightarrow & H_{\mathrm{TP}}^q(\mathfrak{A}) \oplus H_{\mathrm{TP}}^q(\mathfrak{B}) & \rightarrow & H_{\mathrm{TP}}^q(\mathfrak{C}) .
\end{array}$$

The top line is part of a long exact sequence by Mayer–Vietoris. Consequently, the bottom line is part of a long exact sequence. This completes the proof of Theorem 8.11. □

A simplicial map f from K to L preserving the vertex orderings induces maps both in simplicial cohomology and in singular cohomology; one can trace through the isomorphism to see that $\sigma_K^* f_{\mathrm{TP}}^* = f_{\mathrm{PL}}^* \sigma_K^*$ so we have a natural equivalence of functors if we work in the category of simplicial complexes with an ordering on the vertex set.

Let $C_q^\infty(M)$ be the subspace of $C_q^{\mathrm{TP}}(M)$ generated by the piecewise smooth maps of the standard simplex σ_q into M and let $C_\infty^q(M)$ be the dual space. The following fact is proved by standard smoothing arguments; we omit the proof in the interests of brevity.

Lemma 8.12 Let M be a smooth manifold. The inclusion map of $(C_q^\infty(M), \delta)$ into $(C_q^{\mathrm{TP}}(M), \delta)$ and the dual map from $(C_q^{\mathrm{TP}}(M), d)$ to $(C_\infty^q(M), d)$ defines a natural equivalences of functors

$$H_q^\infty(M) \stackrel{\approx}{\longrightarrow} H_q^{\mathrm{TP}}(M) \text{ and } H_{\mathrm{TP}}^q(M) \stackrel{\approx}{\longrightarrow} H_\infty^q(M)\,.$$

Let M and N be compact smooth manifolds without boundary. Let $\omega \in C^\infty(\Lambda^p M)$. We define $\Psi(\omega) \in C_\infty^q(M)$ by defining $\Psi(\omega)$ on the generators. If $f : \sigma_q \to M$ is a piecewise smooth map from the standard q-simplex to M, we define

$$\langle \Psi(\omega), f \rangle = \int_{\sigma_q} (f^*\omega)\,.$$

We apply Stokes' Theorem to see that

$$\begin{aligned} \langle \Psi(d\omega), f \rangle &= \int_{\sigma_q} (f^* d\omega) = \int_{\sigma_q} df^*\omega = \int_{\mathrm{bd}\,\sigma_q} f^*\omega = \langle f^*\omega, \mathrm{bd}\,\sigma_q \rangle \\ &= \langle \omega, \delta f \rangle = \langle df^*\omega, f \rangle\,. \end{aligned}$$

This shows that Ψ is a cochain map from $(C^\infty(\Lambda^P M), d)$ to $(C_\infty^p(M), d)$ and we extend Ψ to a map in de Rham cohomology:

$$\Psi : H_{\mathrm{dR}}^p(M) \to H_\infty^p(M)\,.$$

If ϕ is a smooth map from M to N, then we have a commutative diagram:

$$\begin{array}{ccc} H_{\mathrm{dR}}^p(M) & \stackrel{\Psi_M}{\longrightarrow} & H_\infty^p(M) \\ \uparrow f^* & \circ & \uparrow f^* \\ H_{\mathrm{dR}}^p(N) & \stackrel{\Psi_N}{\longrightarrow} & H_\infty^p(M) \end{array}\,.$$

Consequently, this is a natural transformation of functors. We can now establish a theorem of de Rham [15]. Recall that a finite open cover $\mathcal{U} = \{\mathcal{O}_i\}_{i\in A}$ of a manifold M is said to be a *simple cover* if for any subset B of the indexing set A, the intersection $\cap_{i\in B}\mathcal{O}_i$ is either contractible or empty. By Theorem 5.7, every compact manifold admits a finite simple cover.

Theorem 8.13 *Ψ is a natural equivalence of functors between $H_{\mathrm{dR}}^q(M)$ and $H_\infty^q(M)$ in the category of smooth manifolds without boundary which admit finite simple covers.*

Proof. Suppose M is contractible. Then the homotopy axiom in de Rham cohomology and smooth singular cohomology yields:

$$H^q_{\mathrm{dR}}(M) = H^q_{\mathrm{dR}}(\mathrm{pt}) = H^q_{\infty}(\mathrm{pt}) = H^q_{\infty}(M) = 0 \quad \text{for} \quad q \geq 1\,.$$

On the other hand, $H^0_{\mathrm{dR}}(M) = [1]$ and $H^0_{\infty}(M) = [1_\infty]$ where 1 is the constant 0-form and where $1_\infty(\mathrm{pt}) = 1$ for any point of M. Since $\int_{\{\mathrm{pt}\}} 1 = 1$, the desired isomorphism follows. Note that integration is natural with respect to pullback. Therefore, we have a commutative diagram of short exact sequences in Mayer–Vietoris and hence the connecting homomorphism in Mayer–Vietoris commutes with integration as well. We apply induction over the number of sets in the finite simple cover $\{\mathcal{O}_1, \dots, \mathcal{O}_\ell\}$. Let $\mathcal{U}_1 := \cup_{i<\ell}\mathcal{O}_i$ and let $\mathcal{U}_2 = \mathcal{O}_\ell$. Then $\mathcal{U}_1$ has a finite simple cover with $\ell - 1$ elements and $\mathcal{U}_1 \cap \mathcal{U}_2$ has a finite simple cover with $\ell - 1$ or fewer elements $\{\mathcal{O}_i \cap \mathcal{O}_\ell\}$ for $i < \ell$. We may now apply Mayer–Vietoris (see Theorem 5.2 and Theorem 8.10), the 5-Lemma (see Lemma 8.4), and induction to complete the proof. □

In fact Ψ also preserves the natural ring structures; we shall omit the verification of this fact as defining the natural ring structure on $H^*_\infty(M)$ would take us deeper into the subject than we care to go. Since there is a natural equivalence of functors between $H^*_\infty(\cdot)$ and $H^*_{\mathrm{TP}}(M)$, this identifies de Rham cohomology with topological cohomology with coefficients in $\mathbb{R}$. Topological cohomology can also be defined over $\mathbb{Z}$ and that captures torsion phenomena not present in de Rham cohomology.

Suppose that M is a smooth compact manifold without boundary which is homeomorphic to a finite simplicial complex M. If the homeomorphism is piecewise smooth (i.e., the simplices of K are piecewise smooth submanifolds of M), then the same argument can be used to see that integration gives an isomorphism between $H^*_{\mathrm{PL}}(M)$ and $H^*_{\mathrm{dR}}(M)$. There are "wild triangulations" where the simplicial structure is not smooth. There are also inequivalent PL structures on the same underlying topological manifold. For a further discussion of this and related matters, we refer to the work of Kirby and Siebenmann [35].

8.4 SHEAF COHOMOLOGY

A *sheaf* $\mathcal{F}$ on a topological space X is an assignment $\mathcal{F}(\mathcal{O})$ of a real (or complex) vector space to every open subset $\mathcal{O}$ of X. If $\mathcal{O}$ is the empty set, then we assume $\mathcal{F}(\mathcal{O}) = \{0\}$. If $\mathcal{O}_2$ is an open subset of $\mathcal{O}_1$, we assume given a restriction map $r = r_{\mathcal{O}_2\mathcal{O}_1}$ which is a linear map from $\mathcal{F}(\mathcal{O}_1)$ to $\mathcal{F}(\mathcal{O}_2)$. If $\mathcal{O}_3 \subset \mathcal{O}_2 \subset \mathcal{O}_1$, we assume $r_{\mathcal{O}_3\mathcal{O}_2} r_{\mathcal{O}_2\mathcal{O}_1} = r_{\mathcal{O}_3\mathcal{O}_1}$. Conversely, suppose given $\theta_i \in \mathcal{O}_i$ for $i = 1, 2$. Let $\mathcal{O} = \mathcal{O}_1 \cup \mathcal{O}_2$ and let $\mathcal{O}_{12} = \mathcal{O}_1 \cap \mathcal{O}_2$. If $r_{\mathcal{O}_{12}\mathcal{O}_1}\theta_1 = r_{\mathcal{O}_{12}\mathcal{O}_2}\theta_2$, then there should exist a unique $\theta \in \mathcal{O}$ so $r_{\mathcal{O}_i\mathcal{O}}\theta = \theta_i$ which is called the *extension*. This definition given is a bit complicated and may be safely ignored; it is present just to sooth the mathematical sensibilities of the authors. We shall only need to deal with the following examples.

1. The constant sheaf $\mathbb{R}$. Let $\mathbb{R}(\mathcal{O}) = \mathbb{R}$. The restriction is the identity map.

2. If M is a smooth manifold, let $\mathcal{S}^p$ be the sheaf of smooth p-form; $\mathcal{S}^p(\mathcal{O}) = C^\infty(\Lambda^p(\mathcal{O}))$. The restriction is just the usual restriction of differential forms.

3. If M is a smooth manifold, let $\mathcal{K}^p$ be the sheaf of smooth closed p-forms. This means that $\mathcal{K}^p(\mathcal{O}) = \ker\{d\} \cap C^\infty(\Lambda^p(\mathcal{O}))$. The restriction is again just the usual restriction of differential forms.

It is possible to define sheaf cohomology in complete generality. However, to simplify the discussion and to avoid taking inverse limits, we shall assume that X admits a finite simple open cover; by Theorem 5.7, this is always the case if M is compact manifold. Fix such a cover $\mathcal{U} := \{\mathcal{O}_i\}_{1\le i\le \ell}$ for X henceforth. If $\mathcal{F}$ is a sheaf on X, we may construct a chain complex as follows. Let $I = \{1 \le i_0 < \cdots < i_p \le \ell\}$ be a subset of the indexing set which contains $|I| = p+1$ distinct elements. Let $\mathcal{O}_I := \mathcal{O}_{i_0} \cap \cdots \cap \mathcal{O}_{i_p}$. We consider functions f_p which assign to every such subset I an element of the vector space $\mathcal{F}(\mathcal{O}_I)$; necessarily, of course, $f_p(I) = 0$ if $\mathcal{O}_I$ is empty. Let $C^p(\mathcal{U}, \mathcal{F})$ be the vector space of all such functions where we use the vector space structure on $\mathcal{F}(\mathcal{O}_I)$ to add functions component wise. We define a co-derivative $\delta = \delta_{\mathcal{F}}$ from $C^p(\mathcal{U}, \mathcal{F})$ to $C^{p+1}(\mathcal{U}, \mathcal{F})$ by deleting the indices i_0 through i_{p+1} in succession to define

$$\delta(f_p)(i_0, \ldots, i_p) := \sum_{\nu=0}^{p+1} (-1)^i f_p(i_0, \ldots, \hat{i}_\nu, \ldots, i_{p+1})$$

where we use the restriction map

$$\mathcal{F}(\mathcal{O}_{i_0} \cap \cdots \cap \mathcal{O}_{i_{\nu-1}} \cap \mathcal{O}_{i_{\nu+1}} \cap \cdots \cap \mathcal{O}_{i_p}) \to \mathcal{F}(\cap_\nu \mathcal{O}_{i_\nu})$$

to regard $f_p(i_0, \ldots, \hat{i}_\nu, \ldots, i_{p+1}) \in \mathcal{F}(\mathcal{O}_{i_0} \cap \cdots \cap \mathcal{O}_{i_{p+1}})$. We shall omit this restriction from the notation in the interests of notational simplicity; it is one of the axioms of a sheaf. We compute:

$$\begin{aligned} \delta^2(f_p)(i_0, \ldots, i_{p+2}) &= \sum_{\nu=0}^{p+2} (-1)^\nu \delta(f_p)(i_0, \ldots, \hat{i}_\nu, \ldots i_{p+2}) \\ &= \sum_{\mu<\nu} (-1)^{\mu+\nu} f(i_0, \ldots, \hat{i}_\mu, \ldots, \hat{i}_\nu, \ldots, i_{p+2}) \\ &\quad + \sum_{\mu>\nu} (-1)^{\mu+\nu-1} f(i_0, \ldots, \hat{i}_\nu, \ldots, \hat{i}_\mu, \ldots, i_{p+2}) . \end{aligned}$$

The sign change in the second sum arises as removing i_μ is not the μth index but rather the $\mu^{\text{th}} - 1$ index since we have already removed one of the previous indices. The two sums cancel and we get zero. This is a cochain complex and we denote the cohomology by

$$H^p(\mathcal{U}, \mathcal{F}) := \frac{\ker\{\delta : C^p(\mathcal{U}, \mathcal{F}) \to C^{p+1}(\mathcal{U}, \mathcal{F})\}}{\text{range}\{\delta : C^{p-1}(\mathcal{U}, \mathcal{F}) \to C^p(\mathcal{U}, \mathcal{F})\}} .$$

If $\theta \in C^0(\mathcal{U}, \mathcal{F})$, then $f_i(\theta) \in \mathcal{F}(\mathcal{O}_i)$ for all i. Then $\delta(f) = 0$ if and only if $f_i(\theta) = f_j(\theta)$ on $\mathcal{O}_i \cap \mathcal{O}_j$ or, equivalently, there is a globally defined $\Theta \in \mathcal{F}(M)$ so that $\Theta_{\mathcal{O}_i} = f_i(\theta)$ on $\mathcal{O}_i$ for all i. This shows

$$H^0(\mathcal{U}, \mathcal{F}) = \mathcal{F}(M). \tag{8.4.a}$$

We remark that it is not difficult to show that the sheaf-cohomology groups $H^p(\mathcal{U}, \mathcal{F})$ are in fact independent of $\mathcal{U}$ and, consequently, we obtain $H^p(\mathcal{F})$. But as this would take us a bit deeper into the subject than is necessary, we shall omit the verification of this fact.

We shall now give another proof that de Rham cohomology and topological cohomology coincide by identifying both with sheaf cohomology with coefficients in the constant sheaf $\mathbb{R}$. If M is a compact smooth manifold, we apply Theorem 8.7 to choose a simplicial structure K on M. We fix K henceforth.

Theorem 8.14 Let $M = |K|$ be a smooth manifold which is the realization of a finite simplicial complex K There exists a finite simple open cover $\mathcal{U} = \mathcal{U}(K)$ of M so that:

1. $H^p_{\text{dR}}(M)$ is isomorphic to $H^p(\mathcal{U}, \mathbb{R})$.
2. $H^p(\mathcal{U}, \mathbb{R})$ is isomorphic to $H^p_{\text{PL}}(K)$.
3. $H^p_{\text{PL}}(K)$ is isomorphic to $H^p_{\text{TP}}(M)$.

Proof. Enumerate the vertex set V of K in the form $V = \{v_1, \dots, v_\ell\}$. Let

$$\star(v_i) := \{x \in |K| : t_i(x) > 0\}$$

be the *star of a vertex* v. Let $\mathcal{U} := \{\star(v_1), \dots, \star(v_\ell)\}$ be a finite open cover of $|K|$. If $I \subset V$, the corresponding open set

$$\star(I) := \star(v_{i_0}) \cap \dots \cap \star(v_{i_p}) = \{x \in |K| : t_i(x) > 0 \text{ if } i \in I\}$$

is non-empty if and only if $I \in K$. This set is contractible and deformation retracts to the barycenter $\frac{1}{p+1}\{v_{i_0} + \dots + v_{i_p}\}$ of I. We let $\mathcal{U} = \mathcal{U}(K) := \{\star(v_i)\}$; this is a finite simple cover of M.

Let $\{e_i\}_{1 \le i \le \ell}$ be the standard basis for $\mathbb{R}^\ell$. Let $e^I := e^{i_0} \wedge \dots \wedge e^{i_p}$. Identify f_p in $C^p(\mathcal{U}, \mathcal{S}^q)$ with the sum $\sum_{|I|=p+1} f_p(I) e^I$. We then have

$$\delta(f_p) = \sum_{i=1}^{\ell} \sum_{|I|=p+1} f_p(I) e^i \wedge e^I .$$

The relation $\delta^2 = 0$ is then simply the relation that $\sum_{i,j} e^i \wedge e^j = 0$.

Let $\{\phi_i\}$ be a partition of unity subordinate to $\mathcal{U}$. Since ϕ_{i_ν} vanishes on $\mathcal{O}^c_{i_\nu}$, we may extend $\phi_{i_\nu} f(I)$ to $C^\infty(\Lambda^q(\star(i_0) \cap \dots \cap \star(i_{\nu-1}) \cap \star(i_{\nu+1}) \cap \dots \cap \star(i_p)))$ to be zero on $\star(i_\nu)^c$. This is exactly the construction we used when establishing the Mayer–Vietoris sequence in Theorem 5.2. Let int be interior multiplication, as discussed in Equation (8.1.b). Define

$$\Xi(f_p) := \sum_{\nu=1}^{\ell} \sum_{|I|=p} \phi_i f(I) \operatorname{int}(e^i) e^I \in C^{p-1}(\mathcal{S}^q)\,.$$

Suppose $p > 0$ so that even after deleting an index, there is still an index left. We compute

$$\begin{aligned}
&\operatorname{ext}(e^j)\operatorname{int}(e^i) + \operatorname{int}(e^i)\operatorname{ext}(e^j) = \delta^{ij}\,\mathrm{Id},\\
&\{\Xi\delta + \delta\Xi\} = \sum_{i,j} \phi_i \{\operatorname{ext}(e^j)\operatorname{int}(e^i) + \operatorname{int}(e^i)\operatorname{ext}(e^j)\} = \sum_i \phi_i = \mathrm{Id}\,.
\end{aligned}$$

Therefore, Ξ provides a chain homotopy that shows $H^p(\mathcal{S}^q) = 0$ for $p > 0$; the existence of a "partition of unity" means that $\mathcal{S}$ is a *flabby sheaf.* This proves

$$H^p(\mathcal{U}, \mathcal{S}^q) = 0 \quad \text{if } q \geq 0 \text{ and if } p \geq 1\,. \tag{8.4.b}$$

Let $q > 0$. We have the following short exact sequence:

$$0 \to C^\infty(\Lambda^q(\mathcal{O})) \cap \ker\{d\} \xrightarrow{i} C^\infty(\Lambda^q(\mathcal{O})) \xrightarrow{d} C^\infty(\Lambda^{q+1}(\mathcal{O}_I)) \cap \operatorname{range}\{d\} \to 0\,.$$

Because $\star(I)$ is either empty or contractible, $H^q_{\mathrm{dR}}(\star(I)) = 0$ for $q > 0$. This implies that $\operatorname{range}\{d\} = \ker\{d\}$ so we get a short exact sequence of chain complexes:

$$0 \to C^*(\mathcal{U}, \mathcal{K}^q) \to C^*(\mathcal{U}, \mathcal{S}^q) \to C^*(\mathcal{U}, \mathcal{K}^{q+1}) \to 0 \quad \text{if} \quad q > 0\,.$$

Lemma 8.3 then yields a long exact sequence in cohomology. Equation (8.4.b) implies:

$$0 = H^p(\mathcal{U}, \mathcal{S}^q) \to H^p(\mathcal{U}, \mathcal{K}^{q+1}) \to H^{p+1}(\mathcal{U}, \mathcal{K}^q) \to 0 = H^{p+1}(\mathcal{U}, \mathcal{S}^q) \text{ for } q > 0 \text{ and } p > 0\,.$$

This proves

$$H^{p+1}(\mathcal{U}, \mathcal{K}^q) = H^p(\mathcal{U}, \mathcal{K}^{q+1}) \quad \text{if } q > 0 \text{ and if } p > 0\,. \tag{8.4.c}$$

We use Equation (8.4.a) to see that $H^0(\mathcal{U}, \mathcal{S}^q) = C^\infty(\Lambda^q(M))$. The beginning of the Mayer–Vietoris sequence then yields that:

$$\begin{array}{ccccccc}
C^\infty(\Lambda^p M) & \xrightarrow{d} & C^\infty(\Lambda^p M) \cap \ker\{d\} & & & & 0\\
\downarrow\approx & \circ & \downarrow\approx & & & & \downarrow\approx\\
H^0(\mathcal{U}, S^q) & \to & H^0(\mathcal{U}, \mathcal{K}^{k+1}) & \to & H^1(\mathcal{U}, \mathcal{K}^q) & \to & H^1(\mathcal{U}, \mathcal{S}^q)
\end{array}\,.$$

Consequently, $H^1(\mathcal{U}, \mathcal{K}^q) = H^{q+1}_{\mathrm{dR}}(M)$ and a recursive application of Equation (8.4.c) yields:

$$H^{q+1}_{\mathrm{dR}}(M) = H^1(\mathcal{U}, \mathcal{K}^q) = \cdots = H^{q+1}(\mathcal{U}, \mathcal{K}^0)\,.$$

Since each $\mathcal{O}_I$ is connected, $C^\infty(\Lambda^0(\mathcal{O})) \cap \ker\{d\}$ consists of the constant functions. Therefore, $\mathcal{K}^0$ is the constant sheaf and we have finally $H^p_{\mathrm{dR}}(M)$ is isomorphic to $H^p(\mathcal{U};\mathbb{R})$. This completes the proof of Assertion 1.

It is clear that $C^p(\mathcal{U},\mathbb{R})$ is a vector space with basis the p-simplices of K. Furthermore, the definition of the coboundary operator δ in the sheaf-theoretic context agrees with the coboundary operator in the PL context. Consequently, $H^p(\mathcal{U},\mathbb{R}) = H^p_{\mathrm{PL}}(\mathcal{U})$. Assertion 2 now follows; Assertion 3, which identifies $H^p_{\mathrm{PL}}(\mathcal{U})$ with $H^p_{\mathrm{TP}}(|K|)$, was proved in Theorem 8.11 using the Mayer–Vietoris sequence. □

Bibliography

[1] D. Ado and I. D. Ado, "The representation of Lie algebras by matrices," *Amer. Math. Soc. Translation* **2** (1949); see also "The representation of Lie algebras by matrices," *Uspekhi Mat. Nauk,* **2** (1947), 159–173. 54

[2] A. Arvanitoyeorgos, "An introduction to Lie groups and the geometry of homogeneous spaces," Student Mathematical Library **22**. American Mathematical Society, Providence, R. I. 87

[3] M. F. Atiyah, R. Bott, and A. Shapiro, "Clifford Modules," *Topology* **3** suppl. 1 (1964), 3–38. 17, 29

[4] H. F. Baker, "Alternants and continuous groups," *Proc. London Math. Soc.* **3** (1905), 24–47. 53

[5] S. Bochner, "Curvature and Betti numbers," *Ann. of Math.* **49** (1948), 379–390. 35

[6] J. E. Campbell, "On a law of combination of operators," *Proc. London Math. Soc.* **28** (1897), 381–390. 53

[7] É. Cartan, "Sur la structure des groupes de transformations finis et continus," Thèse présentée à la Faculté des Sciences. 4°. 156 S. Paris. Nony et Co (1894). 71

[8] É. Cartan, "Sur la structure des groupes infinis de transformations," Annales Scientifiques de l'école Normale Supérieure **21** (1904): 153–206. 84

[9] É. Cartan, "Sur une classe remarquable d'espaces de Riemann, I," *Bull. Soc. Math. France* **54** (1926), 214–216. 102

[10] É. Cartan, "Sur une classe remarquable d'espaces de Riemann, II," *Bull. Soc. Math. France* **55** (1927), 114–134. 102

[11] É. Cartan, "Les représentations linéaires des groupes de Lie," *J. Math. Pures Appl.* **IX** Sér. 17 (1938), 1–12. 54

[12] H. Cartan and S. Eilenberg, "Homological algebra," Princeton University Press, Princeton, N.J. (1956). 114

[13] A. Cauchy, "Mémoire sur les intégrales définies," Oeuvres complètes Ser. 1 1, Paris (published 1882), 319–506. 9

[14] C. Chevalley and S. Eilenberg, "Cohomology theory of Lie groups and Lie algebras," *Trans. Amer. Math. Soc.* **63** (1948), 85–124. 81

[15] G. de Rham, "La théorie des formes différentielles extérieures et l'homologie des variétés différentiables," *Rend. Mat. e Appl.* **20** (1961), 105–146. 127

[16] C. DeWitt-Morette and B. DeWitt, "Pin groups in physics," *Phys. Rev. D* **41** (1990), 1901–1907. 54

[17] S. Eilenberg and N. Steenrod, "Axiomatic approach to homology theory," *Proc. Nat. Acad. Sci. U. S. A.* **31** (1945), 117–120. 17, 118, 125

[18] S. Eilenberg and N. Steenrod, "Foundations of algebraic topology," Princeton University Press, Princeton, N.J. (1952). 17, 116, 118, 125

[19] J. Fourier, "Mémoire sur la propagation de la chaleur dans les corps solides," 215–221 Présenté le 21 décembre 1807 à l'Institut national – *Nouveau Bulletin des sciences par la Société philomatique de Paris*, t. I, 112–116, n¡6; mars 1808. Paris, Bernard. 32

[20] P. Gilkey, "Invariance theory, the heat equation, and the Atiyah–Singer index theorem," *Studies in Advanced Mathematics*, CRC Press, Boca Raton (1995). 31, 44

[21] P. Gilkey, C. Y. Kim, H. Matsuda, J. H. Park, and S. Yorozu, "Non-closed curves in $\mathbb{R}^n$ with finite total first curvature arising from the solutions of an ODE," to appear *Hokkaido Mathematics Journal.* 1, 2

[22] P. Gilkey, C. Y. Kim, and J. H. Park, "Real analytic complete non-compact surfaces in $\mathbb{R}^n$ with finite total curvature arising as solutions to ODEs," to appear *Tohoku Mathematics Journal.* 1, 3

[23] A. Gray, "The volume of a small geodesic ball of a Riemannian manifold," *Michigan Math. J.* **20** (1974), 329–344. 8

[24] M. Guediri, "On completeness of left-invariant Lorentz metrics on solvable Lie groups," *Rev. Mat. Univ. Complut. Madrid* **9** (1996), 337–350. 101

[25] M. Guediri and J. Lafontaine, "Sur la complétude des variétés pseudo-riemanniennes," *J. Geom. Phys.* **15** (1995), 150–158. 101

[26] B. Hall, "Lie groups, Lie algebras, and representations. An elementary introduction," *Graduate Texts in Mathematics* **222**, Springer–Verlag, New York (2003). 59, 60

[27] G. S. Hall, "The global extension of local symmetries in general relativity," *Class. Quantum Grav.* **6** (1989), 157–161. 99

[28] Harish-Chandra (1949), "Faithful representations of Lie algebras," *Ann. of Math.* **50** (1949), 68–76. 54

[29] F. Hausdorff, "Die symbolische Exponentialformel in der Gruppentheorie," *Leipz. Ber.* **58** (1906), 19–48. 53

[30] S. Helgason, "Differential geometry and symmetric spaces," *Pure and Applied Mathematics, vol. XII*, Academic Press, New York–London (1962). 60, 87, 109

[31] S. Helgason, "Differential geometry, Lie groups, and symmetric spaces," *Pure and Applied Mathematics* **80**. Academic Press, Inc. [Harcourt Brace Jovanovich, Publishers], New York–London (1978). 60

[32] W. V. D. Hodge, "The Theory and Applications of Harmonic Integrals," Cambridge University Press, Cambridge (1941). 34

[33] H. Hopf, "Über die Topologie der Gruppen-Mannigfaltigkeiten und ihre Verallgemeinerungen," *Ann. Math.* **42** (1941), 22–52. 82

[34] W. Hurewicz, "On duality theorems (abstract 47-7-329)," *Bull. of the Amer. Math. Soc.* **47** (1941), 562–563. 113

[35] R. C. Kirby and L. Siebenmann, "Foundational Essays on Topological Manifolds, Smoothings, and Triangulations," Princeton University Press, Princeton, N.J. (1977). 128

[36] A. Kirillov, "Introduction to Lie groups and Lie algebras," *Cambridge Studies in Advanced Mathematics* **113**, Cambridge University Press, Cambridge (2008). 45, 54

[37] S. Kobayashi, "Theory of connections," *Ann. Mat. Pura Appl.* **43** (1957), 119–194. 99

[38] H. Künneth, "Über die Torsionszahlen von Produktmannigfaltigkeiten," *Math. Ann.* **91** (1924), 125–134. 39

[39] J. Marsden, "On completeness of homogeneous pseudo-riemannian manifolds," *Indiana Univ. J.* **22** (1972/73), 1065–1066. 100

[40] W. Massey, "A basic course in algebraic topology," *Graduate texts in mathematics* **127**, Springer–Verlag, New York (1991). 113, 115

[41] W. Mayer, "Über abstrakte Topologie," *Monatshefte für Mathematik* **36** (1929), 1–42. 20, 119, 125

[42] K. Melnick, "Isometric actions of Heisenberg groups on compact Lorentz manifolds," *Geom. Dedicata* **126** (2007), 131–154. 94

[43] C. Meneghini, "Clifton-Pohl torus and geodesic completeness by a 'complex' point of view," *Complex Var. Theory Appl.* **49** (2004), 833–836. 101

[44] W. Miller, "Symmetry Groups and their Applications," Academic Press, New York (1972). 53

[45] S. B. Myers and N. E. Steenrod, "The group of isometries of a Riemannian manifold," *Ann. of Math.* **40** (1939), 400–416. 93

[46] A. Newlander and L. Nirenberg, "Complex analytic coordinates in almost complex manifolds," *Ann. of Math.* **65** (1957), 391–404. 9

[47] A. Nijenhuis and W. Woolf, "Some integration problems in almost-complex and complex manifolds," *Ann. of Math.* **77** (1963), 424–489. 9

[48] K. Nomizu, "On local and global existence of Killing vector fields," *Ann. Math.* **72** (1960), 105–120. 99

[49] B. O'Neill, "Semi-Riemannian geometry. With applications to relativity," *Pure and Applied Mathematics* **103**, Academic Press, Inc. (1983), New York. 87

[50] R. Palais, "A global formulation of the Lie theory of transformation groups," *Mem. Amer. Math. Soc.* **22** (1957). 93

[51] J. Patera, R.T. Sharp, P. Winternitz, and H. Zassenhaus, "Invariants of real low dimension Lie algebras," *J. Mathematical Phys.* **17** (1976), 986–994. 49

[52] F. Peter and H. Weyl, "Die Vollständigkeit der primitiven Darstellungen einer geschlossenen kontinuierlichen Gruppe," *Math. Ann.* **97** (1927), 737–755. 68

[53] H. Poincaré, *Analysis Situs, J. de l'Éc. Pol* **I** (1895), 1–123. 37

[54] B. Riemann, "Grundlagen für eine allgemeine Theorie der Funktionen einer veränderlichen komplexen Grösse,"(1851), in H. Weber, Riemann's gesammelte math. Werke, Dover (published 1953), 3–48. 9

[55] R. Saerens and W. Zame, "The isometry groups of manifolds and the automorphism groups of domains," *Trans. Amer. Math. Soc.* **301** (1987), 413–429. 93

[56] E. Spanier, "Algebraic topology," Springer–Verlag (1995), Berlin. 125

[57] M. Spivak, "A comprehensive introduction to differential geometry," Second edition. Publish or Perish Press (1979), Wilmington, Del. 57

[58] W. T. van Est, "Une démonstration de E. Cartan du troisieme theoreme de Lie." *Action hamiltoniennes de groupes. Troisieme theoreme de Lie* (Lyon, 1986), 83–96, Travaux en Cours, **27**, Hermann (1988), Paris. 54

[59] L. Vietoris, “Über die Homologiegruppen der Vereinigung zweier Komplexe,” *Monatshefte für Mathematik* **37** (1930), 159–162. 20, 119, 125

[60] F. Warner, “Foundations of differentiable manifolds and Lie groups,” Scott, Foresman and Co. (1971), Glenview, Ill.-London. 87

[61] R. Weitzenböck, “Invariantentheorie,” Groningen, Noordhoff, 1923. 35

[62] H. Whitney, “Differentiable manifolds,” *Ann. of Math.* **37** (1936), 645–680. 121

[63] H. Whitney, “Geometric integration theory,” Princeton University Press (1957), Princeton N.J. 121

[64] W. Ziller, “Lie groups, representation theory and symmetric spaces,” Notes from courses taught Fall 2010 at Penn and 2012 at IMPA, published electronically http://www.math.upenn.edu/ wziller/math650/LieGroupsReps.pdf 87

Authors' Biographies

PETER B GILKEY

Peter B Gilkey[1] is a Professor of Mathematics and a member of the Institute of Theoretical Science at the University of Oregon. He is a fellow of the American Mathematical Society and is a member of the editorial board of Results in Mathematics, J. Differential Geometry and Applications, and J. Geometric Analysis. He received his Ph.D. in 1972 from Harvard University under the direction of L. Nirenberg. His research specialties are Differential Geometry, Elliptic Partial Differential Equations, and Algebraic topology. He has published more than 250 research articles and books.

JEONGHYEONG PARK

JeongHyeong Park[2] is a Professor of Mathematics at Sungkyunkwan University and is an associate member of the KIAS (Korea). She received her Ph.D. in 1990 from Kanazawa University in Japan under the direction of H. Kitahara. Her research specialties are spectral geome-try of Riemannian submersion and geometric structures on manifolds like eta-Einstein manifolds and H-contact manifolds. She organized the geometry section of AMC 2013 (The Asian Mathematical Conference 2013) and the ICM 2014 satellite conference on Geometric analysis. She has published more than 71 re-search articles and books.

[1]Mathematics Department, University of Oregon, Eugene OR 97403 U.S.
email: `gilkey@uoregon.edu`
[2]Mathematics Department, Sungkyunkwan University, Suwon, 440-746, Korea
email: `parkj@skku.edu`

RAMÓN VÁZQUEZ-LORENZO

Ramón Vázquez-Lorenzo[3] is a member of the research group in Riemannian Geometry at the Department of Geometry and Topology of the University of Santiago de Compostela (Spain). He is a member of the Spanish Research Network on Relativity and Gravitation. He received his Ph.D. in 1997 from the University of Santiago de Compostela under the direction of E. García-Río. His research focuses mainly on Differential Geometry with special emphasis on the study of the curvature and the algebraic properties of curvature operators in the Lorentzian and in the higher signature settings. He has published more than 50 research articles and books.

[3]Department of Geometry and Topology, Faculty of Mathematics, University of Santiago de Compostela, 15782 Santiago de Compostela, Spain.
email: ravazlor@edu.xunta.es

Index